Free Energy's Flow: Nature's Engine

Naina

Contents

1

Introduction

Thermodynamic systems strive towards a state of minimal free energy. As such, the free energy — as well as changes therein — is a central quantity for understanding and predicting a wide variety of phenomena in solid state, soft matter and biophysics [1–7], and, in extension, useful to a range of biomedical applications [8–11] or material design [12–15]. For example, predicting the folded structure of a protein requires finding its free energy minimum. These predictions remain a long-standing problem in biology and would help address, e.g., neurodegenerative diseases caused by misfolding and subsequent protein aggregation [16–18].

Along similar lines, free energy differences determine how strongly and in which likely conformation a (potential) ligand binds to a protein [19, 20], thereby enacting signaling paths which comprise the communication between cells. In pathological cases, knowledge thereof aids the design of ligand-based drugs, which, for example, modulate the actions of a receptor [11, 21]. Furthermore, the rate at which, e.g., a ligand reversibly interchanges between a bound and an unbound state is directly related to the energetic barriers separating the two states [22, 23]. In addition, the difference in solvation free energy, which will also be addressed in this thesis, is related to the partition coefficient of a solute between organic and aqueous solutions. These coefficients provide a measure of hydrophobicity, thereby indicating the ability of, e.g., a drug candidate to permeate through a cellular lipid membrane [24, 25] and reach its target.

Differences in free energies can directly or indirectly be inferred from experiments. For example, binding free energies are obtained via isothermal titration calorimetry [26, 27] by measuring the heat exchange during the binding process. Free energy differences between protein conformations are inferred from the relative amount that they are encountered in, e.g., solution state nuclear magnetic resonance (NMR) [28, 29]. Furthermore, optical tweezers [30, 31] and single-molecule atomic force microscopy (AFM) [32–34] experiments deduce folding free energies from the work required to pull a protein from a folded to an unfolded state. Partition coefficients are determined, among other techniques, using high performance liquid chromatography (HPLC) [35–37] or UV-Vis Spectroscopy [38].

However, obtaining such information for a large variety of molecules, as, e.g., routinely required in "hit to lead" and "lead optimization" stages of drug development processes, is very costly due to the complexity of synthesizing a large number of different compounds. Hence, there is a large interest in accurate computational predictions. In addition, some of the employed models and techniques give further mechanistic insights at a molecular level that are inaccessible via experiments due to temporal and spacial resolution limits.

Computational approaches to predict free energies can broadly be divided in "data driven" and "first principles" approaches. The former category includes, e.g., statistical inference, machine learning (ML) and artificial intelligence (AI) methods. These approaches identify correlations in an underlying training data set and extract the most relevant features for accurate predictions. Examples include estimating binding affinities based on structural elements [39, 40] or using neural networks for predicting partition coefficients depending on the atomic or amino acid composition of a protein [41, 42]. The main advantage of these approaches is their relatively small computational effort once properly trained. Their main disadvantage is their need for large training data sets, as well as their inability to correctly predict free energies for molecules that lack resemblance to those in the training set. On an interesting side note, one of the major models that describes the perceptual processes of a neuronal net and the brain — that underly some of the above data driven methods — also follows a "free energy principle" [43–46]. Whereas this principle does not refer to the minimization of an actual physical energy, it is based on the same probabilistic framework and constructs from statistical physics as the ones used in this thesis.

In contrast, first principles approaches are based on physically motivated models. For example, docking methods fit a compound to a target by minimizing interaction energies using force fields that have been developed using either experimental measurements [47, 48] or *ab initio* calculations [49, 50]. In the latter case, conceptually, these models are derived solely from theory and do not require any experimental data to make predictions for novel compounds. However, whereas docking methods mainly rely on optimizing the enthalpic contributions, the free energy additionally depends on entropic effects. As such, sampling based approaches, such as Monte Carlo (MC) or Molecular Dynamics (MD) simulations, can be used, where the latter integrates the classical equations of motion while representing the system on an atomistic basis. Generally, the methodology depends on the system scale and required level of detail. For larger systems, coarse-grained simulations [51, 52] that approximate several atoms or chemical groups as one particle can be used to analyze the free energies of, e.g., RNA folding [53]. In contrast, on a smaller scale where quantum mechanical (QM) effects cannot be solely approximated through force fields anymore, sampling with *ab initio* methods [54–56] is used to study, e.g., the free energy landscape associated with reactive sites of an enzyme [57]. Naturally, it can also be advantageous to combine data driven with first principles approaches. For example, ML can be used to determine force field parameters [58, 59]. Conversely, MD simulations can be used to generate data that ML approaches are trained on and physical constraints can be incorporated into ML methods [60].

MD simulations, which are used in this work for sampling, have become increasingly popular and the ground work was recognized by the nobel prize in chemistry to Martin Karplus, Michael Levitt and Arieh Warshel in 2013 [61–63]. The widespread use of MD has been further supported by the increased availability of structures [64, 65], as well as progress in computing power, especially for graphical processing units (GPU) that are well suitable for the calculation of interaction forces [66–68]. For MD based free energy calculations, the accuracy of their predictions is influenced by two main factors: Firstly, and similar to most MD applications, the quality of the force fields [69, 70]. Secondly, and most relevant for this work, the sampling approach. For instance, in the recent SAMPL6 blind challenge [71], the participants were asked to calculate a range of binding free energies based on their sampling approach of choice, but using only MD with force field parameters that had been provided by the organizers. The outcome demonstrated that the free energy predictions substantially differ between the individual sampling approaches.

In terms of sampling, the challenge in making accurate predictions lies in the fundamental characteristic of the free energy

$$F = -\beta^{-1} \ln \left(\frac{1}{h^3} \int_{-\infty}^{\infty} \int_{-\infty}^{\infty} e^{-\beta H(\mathbf{x},\mathbf{p})} \mathrm{d}\mathbf{x}\,\mathrm{d}\mathbf{p} \right) \tag{1.1}$$

itself, where the thermodynamic beta is denoted by $\beta = (k_B T)^{-1}$, k_B the Boltzmann constant, T the temperature and h the Planck constant. The positions and momenta of all particles are denoted by $\mathbf{x}$ and $\mathbf{p}$, respectively. A thermodynamic system does not rest in a single micro state $\{\mathbf{x}, \mathbf{p}\}$ the entire time, but instead is constantly changing, and the probability of encountering each micro state decreases exponentially with the Hamiltonian $H(\mathbf{x}, \mathbf{p})$, which is for the cases in this thesis (i.e., absence of external potentials, zero center-of-mass movement and including correction terms for changes in volume) the enthalpy of the system associated with $\mathbf{x}$ and $\mathbf{p}$. For example, a specific configuration of an unbound ligand may be much less favorable in enthalpy than a bound one, and as such, less likely. However, the amount of unbound configurations — or, in continuum, the configuration space volume — is far greater in the first case. Therefore, the unbound state, considered as the combination of all micro states that belong to this category, has a much higher entropy. The free energy, expressed in macroscopic variables

$$F = H - TS \; , \tag{1.2}$$

combines the contributions of enthalpy H and (temperature weighted) entropy S and indicates which state is the more likely one (note that in a constant pressure regime it is referred to as the Gibbs free energy and will be denoted as G, for further definitions see section 2.1).

Consequently, determining the exact free energy requires considering all theoretically possible micro states, a number that is astronomically high for many of the applications mentioned above, and is, therefore, not feasible for sampling based *in silico* approaches. However, in almost all of these applications, it suffices to know only the free energy difference between various states, e.g., between a molecule either in an aqueous or an organic solution. In contrast, knowledge of the absolute free energies, such as the one in aqueous solution alone, is not required. Fortunately, as outlined below, estimates of these differences can often be obtained much more efficiently, and therefore more accurately, than of the absolute ones.

For sampling based approaches, some of the underlying principles can be illustrated by using the analogy of a dart board, as shown in Fig. 1.1. The task consists of obtaining the difference in area between two forms, such as a pentagon and a hexagon (blue and red, respectively), representing, for example, the configuration space volumes of two different ligands. Sampling is conducted by a beginner player randomly throwing darts onto a board (grey circle), leaving the green holes behind.

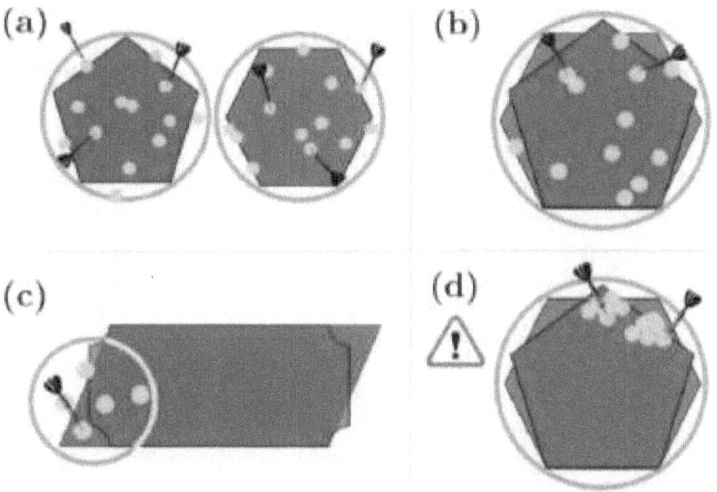

Figure 1.1: "Dart board sampling". The (difference in) area between two forms (blue, red) is determined by throwing darts at random onto a board (grey circle). The green dots indicate the resulting holes. (a) Separate sampling. (b) Simultaneous sampling, assuming the forms can be taken off the board to count the holes in each one of them in the end. (c) Importance sampling, i.e., sampling only within regions relevant to the differences. (d) Unequal sampling, which is undesired. The dart player has lost the ability to sample the entire board, but instead only throws at the upper right area.

The area is obtained as a fraction of the circle by counting the number of holes inside the respective shape compared to the overall number of holes. The accuracy of this approach can be evaluated by comparing how much an estimate based on, e.g., one hundred holes deviates on average from the exact area.

Several approaches to obtain the difference in area can be distinguished. To start with, the area of each shape can be determined separately (panel a), and subtraction yields the difference. Assuming that it takes much more time to throw the darts compared to counting the resulting holes inside the two sheets pinned to the board, efficiency can be gained by throwing the darts onto both forms at once (panel b). For atomistic simulations, the same principle applies, as it is much more difficult to obtain statistically uncorrelated samples compared to evaluating the energies of these under different conditions. Furthermore, efficiency can be gained through importance sampling: In this case, by directing the samples to regions most relevant for the difference (panel c). The concept of intermediate states, which will be introduced below, falls into this category. Lastly, complications arise if samples are not distributed equally over the dart board (panel d), representing, e.g., the complication of an MD simulation to exit a local energy minimum. Even if the increased probability to obtain samples in the upper right corner of the shapes is accounted for in post processing, the result would be more accurate if the sample points were more evenly distributed. Naturally, the dart board analogy is simplified in a number of ways. One of them is that samples can only have values of zero (outside) and one (inside), whereas the (Boltzmann weighted) Hamiltonians can reach a large range of values.

In analogy to the overlapping shapes in the dart board example, calculating the differences in free energy becomes more accurate the more similar the configuration space densities of the considered states are. As such, whereas, e.g., the binding free energy is also only a difference between a bound and an unbound ligand, the entirely different configurations still make it a difficult task. It became possible in the recent decade to calculate binding free energies with an accuracy of below 1 kcal/mol [72–75], however, considerable computing power is still required. Their advantage is that they represent a criterion similar to a scoring function, which can be used for data base screening. However, it is often sufficient to know only which out of two or several ligands binds more favorably, i.e., calculating only relative binding energies (therefore, technically, the difference of a difference).

This case is an example in which thermodynamic cycles can be used, as are illustrated in Fig. 1.2. Here, the horizontal arrows indicate the two absolute binding free energies, ΔG^A_{bind} and ΔG^B_{bind} of ligands A and B, respectively. These are, however, not calculated directly. Instead, due to the higher similarity in configuration space density, a higher accuracy is achieved by calculating the differences along the vertical arrows [10, 76], i.e., between the two receptor bound ligands, $\Delta G^{AB}_{\text{receptor}}$, as well as between the unbound states, where the latter is identical to the difference in solvation free energy $\Delta G^{AB}_{\text{solv}}$. Using the fact that the free energy is a thermodynamic state function and that the differences along the entire cycle sum up to zero, the relative binding free energy $\Delta\Delta G_{\text{bind}}$ is obtained.

Furthermore, when comparing two ligands in solution as required for $\Delta G^{AB}_{\text{solv}}$, the configuration space densities may still not be similar enough. The accuracy can

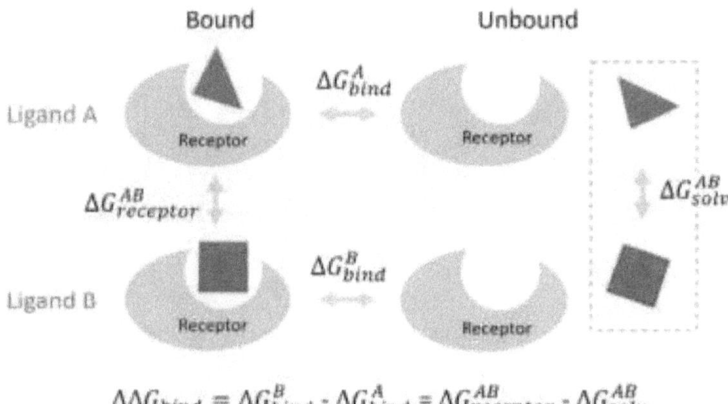

$$\Delta\Delta G_{bind} = \Delta G_{bind}^{B} - \Delta G_{bind}^{A} = \Delta G_{receptor}^{AB} - \Delta G_{solv}^{AB}$$

Figure 1.2: Thermodynamic cycle to calculate relative binding free energies. Two ligands A and B are depicted by a red triangle and a blue square, respectively, that bind to a receptor (grey). Instead of calculating the binding free energies of A and B directly (horizontal green arrows), for reasons of accuracy the differences between the two ligands in the bound state and in the unbound states are determined (vertical arrows), where the latter corresponds to the difference in solvation free energy. As the energies in the circle sum up to zero, the vertical calculations also yield the relative binding free energy, and thereby the desired information as to which ligand binds more favorably.

be further improved through bridge sampling [77, 78], a specific form of importance sampling that is illustrated in Fig. 1.1(c). Rather than comparing the states of interest directly, sampling is conducted in intermediate states that are defined as a function of the Hamiltonians $H_A(\mathbf{x}, \mathbf{p})$ and $H_B(\mathbf{x}, \mathbf{p})$ that define the ligands A and B. Most commonly, these are interpolated linearly

$$H_\lambda(\mathbf{x}, \mathbf{p}) = (1 - \lambda)H_A(\mathbf{x}, \mathbf{p}) + \lambda H_B(\mathbf{x}, \mathbf{p}) \ , \tag{1.3}$$

where λ denotes the path variable. For the example of differing geometric shapes, Fig. 1.3(a) illustrates how the forms of the end states A and B are not compared directly, but rather through a morphing sequence of stepwise changing ones. From the pairwise differences between neighboring shapes, the overall difference is obtained. For two Hamiltonians consisting of harmonic potentials with

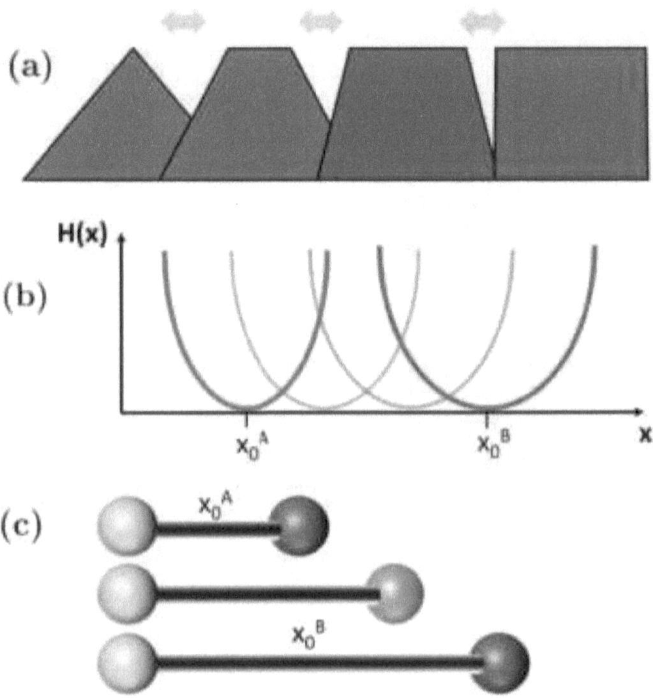

Figure 1.3: Alchemical transformations. To calculate the, e.g., solvation free energy difference between two ligands indicated by the dashed box in Fig. 1.2, intermediate states are used. (a) Geometric illustration of how the shapes gradually change between the red and blue ones that the differences is to be calculated of. The green arrows indicate that the differences are calculated between neighboring shapes. (b) Linear interpolation series of intermediate Hamiltonians (transparent lines) between two harmonic potentials with different rest length x_0^A and x_0^B, as well as different spring constants. (c) Chemical bonds between two atoms, where the type of the right one differs between, e.g., the two ligands of interest. As a consequence, the rest length changes (shown exaggeratedly). For an intermediate state, the linearly interpolated Hamiltonian can be interpreted as simulating with an atom type (grey) with a rest length between x_0^A and x_0^B.

minima at x_0^A and x_0^B, shown in Fig. 1.3(b), the Hamiltonians determined through the linear interpolation scheme (transparent colors), Eq. 1.3, govern the intermediate states.

The use of such a morphing sequence or path is referred to as alchemical transformation. The term "alchemical" reflects the fact that if a number of atoms differ between A and B, then these atoms are transformed into one another, thereby realizing the proverbial alchemical dream of transforming matter. Hence, also the thermodynamic cycle shown in Fig. 1.2 is sometimes referred to as an alchemical cycle. As an example, the interpolated Hamiltonian shown in Fig. 1.3(b) also has the form of a harmonic potential. Therefore, for a bond where one of the two atoms changes its type, as illustrated by the red and blue colors in Fig. 1.3(c), the intermediate state can be interpreted as an atom type (grey) with a bond rest length between the one belonging to the end states. Even though such intermediate atom types do not exist in physical reality, the information gained from sampling in these intermediate states considerably improves the accuracy of the free energy estimates. In case the ligands differ by entire chemical groups with, e.g., ligand A having a larger number of atoms than ligand B, then excess atoms are transformed into "ghost" particles with an identical mass, but zero interaction energies in B to ensure equal dimensionality of $\mathbf{x}$ between A and B. These will be referred to as "vanishing particles", and the process as "decoupling" in this thesis.

Two types of approaches can be distinguished to calculate free energy differences along such an alchemical transformation: Firstly, for equilibrium sampling, simulations are conducted at discrete steps of the path variable λ on time scales for which, as the name implies, it can be assumed that samples are randomly drawn from the configuration space density corresponding to thermal equilibrium. Estimators have been developed in the context of Free Energy Perturbation (FEP) theory, such as the Zwanzig formula [79] or the Bennett Acceptance Ratio (BAR) method [80], to evaluate the difference between adjacent intermediates. Secondly, for non-equilibrium approaches, λ is continuously changed in each integration step within a simulation, that is, therefore, out of equilibrium. Initially, this approach built the basis of the slow-growth method [81–83], where λ is increased so slowly that the system is nonetheless assumed to be close enough to equilibrium, such that the required work to transform the system between A and B equals the free energy difference. Later, Jarzynski [84] laid the theoretical foundations that relate these work values with the free energy difference — even arbitrarily far from equilibrium — therefore also allowing to increase λ much faster. As a consequence, non-equilibrium approaches became more popular and widely used to calculate free energy differences in various contexts [85–89]. A special case between the two types of approaches is Thermodynamic Integration (TI) [90], which has been derived assuming continuous sampling along λ. However, sampling is nonetheless conducted at equilibrium in a number of states similar to the ones used for FEP. The free energy difference between adjacent states is then determined through numerical integration over λ.

A number of alterations to the linear interpolation scheme, Eq. 1.3, have been developed, with soft-core methods [91–96] being the most prominent ones. For simulations of systems with vanishing particle, these will generally overlap with other ones in one of the end states. When calculating the energy of such configurations at states where the particle is still interacting, divergences occur for Lennard-Jones (LJ) and Coulomb interactions. Therefore, various λ-dependences of $H_A(\mathbf{x}, \mathbf{p}, \lambda)$ and $H_B(\mathbf{x}, \mathbf{p}, \lambda)$ were constructed such that these divergences are avoided. Further alternatives to Eq. 1.3 are the One-Step Perturbation (OSP) [97–99] and Enveloping Distribution Sampling (EDS) [100–103] methods. These methods calculate the difference between several end states by simulating in a reference potential that "envelopes" their configuration space densities, i.e., that combines all regions in configuration space relevant to at least one of the end states. Furthermore, for TI, the Minimum Variance Path (MVP) [78, 104, 105] has been derived that interpolates the configuration state densities rather than the Hamiltonians along λ.

A different class of methods also alters the free energy landscape, but aims at alleviating the frequent complication that not all relevant regions in configuration space are sufficiently sampled, as also illustrated for the dart board example in Fig. 1.1(d). These complications arise because the free energy landscapes of biomolecules are generally rugged [106–109], and therefore energetic barriers are not crossed at sufficiently high rates.

To address this problem, two types of methods can be distinguished: Firstly, for a given state, biasing methods gradually raise the energy levels for configurations that have already been sampled, thereby forcing the system to evolve into other previously less sampled regions. Variants that rely on these principles are, for example, conformational flooding [110, 111], metadynamics [112–115] or accelerated MD [116–119].

Secondly, Replica Exchange (RE) methods conduct several simulations in parallel and switch configurations between these using a Metropolis criterion [120] to sample different regions of configuration space. For Hamiltonian RE [121–124] methods, these simulations are conducted in states such as the ones defined by the linear and soft-core interpolation schemes, Eq. 1.3, with different λ-values. RE with solute tempering [125–127] further deforms the Hamiltonian as such that the acceptance probability for the exchange of replica configurations is increased, yielding an improved efficiency. For Parallel Tempering [128–132] (the name being often used synonymously to RE) simulations are conducted at different temperatures of the state of interest and above, thereby increasing the variety of configurations that are exchanged.

In addition to determining free energy differences between two states, variants of the above techniques can also be used to calculate the free energy along an actual physical path, thereby providing information on barriers along this path. For example, the linear alchemical path for calculating the free energy difference between an ion located on the inside and the outside of a cellular membrane would mean that the ion interactions on the inside would be gradually decoupled, while oppositely, a decoupled particle on the outside would be transformed into an ion. Whereas the alchemical path is likely to predict the free energy difference more accurately than a physical one, it neither reveals the permeation rate nor the mechanics of the corresponding ion channel. The most intuitive approach to this aim is free sampling, where the free energy difference is obtained based on spontaneous transitions of the system between the end states. However, especially for high energetic barriers along the path, this approach may yield only poor statistics.

In this case, as an example of an equilibrium approach, Umbrella Sampling (US) [133–135] is the method of choice. In the first step, a finite number of points along the physical path is selected, in this case through the ion channel. In the next step, harmonic potentials are applied to restrain the sampling region around this point. These "umbrella windows" have to be chosen such that the resulting configurational distributions have sufficient overlap. The free energy along the path is then obtained via the Weighted Histogram Analysis Method (WHAM) [136–138] that counts occurrences of configurations in bins and accounts for the umbrella bias. Without binning (i.e., in the limit of zero-bin width), WHAM reduces to the Multistate Bennett Acceptance Ratio (MBAR) method, a generalization of BAR that is also used between alchemical states. WHAM and BAR can also be derived from a unifying Bayesian approach [139].

An example for a similar non-equilibrium approach along a physical path is Steered Molecular Dynamics (SMD) [140–142], where either a constant force is applied to one or several atoms, or these atoms are moved with a constant velocity while keeping track of the required force or work to do so. In this sense, SMD mimics AFM experiments. Among other purposes, SMD can, in combination with Jarzynski's theory [84], be used to create free energy profiles [143, 144].

Lastly, in cases where such a path (or multiple ones) between two states are of interest but unknown, transition path finding methods are used. These generally use an initial trial path that is varied to determine the optimum. For example, chain-of-state methods, such as the Nudged Elastic Band [145–147] or the String method [148–151] aim at finding the minimum energy path. Other methods, such as action-derived MD [152–154] are based on the principle of least action. Extensions, such as the Action Conformational Space Annealing method [155, 156], penalize searches too close to the initial trial path to avoid identifying only the closest local minimum.

This thesis focuses on the theory of alchemical transformations to determine free energy differences. For these transformations, the most commonly used path is the linear interpolation scheme between the end state Hamiltonians, Eq. 1.3, and soft-core variants thereof. This path is illustrated by the straight arrow in Fig. 1.4. However, the linear interpolation is only a very special case among all possible ways to change $H_A(\mathbf{x}, \mathbf{p})$ into $H_B(\mathbf{x}, \mathbf{p})$, as indicated by the curved arrows. Only few have so far been considered in the literature and, with the exception of the MVP, have mostly been empirically constructed based on a trial and error optimization.

The present thesis therefore addresses the question: From all possible transformations, which sequence of intermediate states yields free energy estimates with optimal accuracy?

Chapter 2 will give an introduction about the fundamentals of the free energy methods relevant for this work, as well as the underlying principles of MD simulations.

In chapter 3, for FEP and BAR, the sequence of intermediate states that yields estimates of free energy differences with minimal mean squared error (MSE) will be derived using variational calculus. These resulting states will be referred to

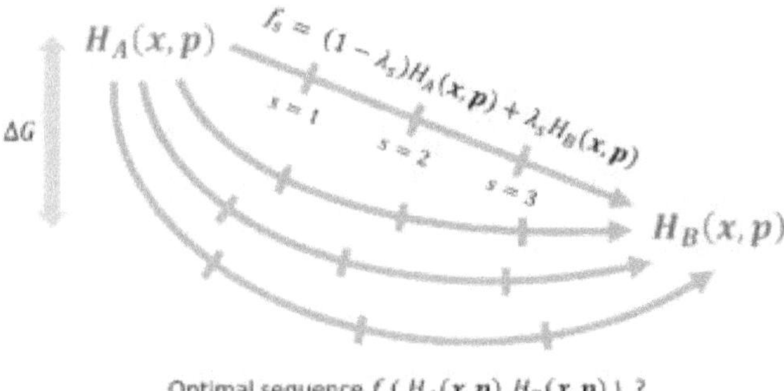

Figure 1.4: Question of this thesis. Different paths connecting the two end state Hamiltonians, $H_A(\mathbf{x}, \mathbf{p})$ and $H_B(\mathbf{x}, \mathbf{p})$, are possible. Most commonly, a linear interpolation is employed (straight upper grey arrow); however, alternative ones (curved arrows below) are also possible. The grey bars denote discrete states labeled by s that simulations are conducted in for alchemical equilibrium techniques. This thesis addresses the question of the functional form f_s defining the optimal sequence of intermediate states.

as Variationally Derived Intermediates (VI) and compared against existing methods for one-dimensional test systems. A parallelizable approximation to the VI sequence will be suggested and assessed through test systems of increasing complexity, which consist of a 1-D potential, a LJ gas and the electrostatic decoupling of butanol in solution.

In chapter 4, the derivation of a more efficient variant of VI, correlated Variationally Derived Intermediates (cVI), will be developed and assessed. It considers the common practice to increase efficiency by using the same sample points from an intermediate state to evaluate the difference to both adjacent ones. For this practice, however, correlations arise between the step-wise free energy estimates that had not been considered up to this point. The cVI sequence yields the optimal MSE under these conditions.

The implementation of the approximated VI sequence suggested in chapter 3 into the Groningen Machine for Chemical Simulations (GROMACS) MD software package [157–159] will be described in detail in chapter 5. In addition, the method will be extended by an approach to avoid numerical instabilities for vanishing particles. An assessment of its accuracy will be provided, as well as example cases that describe the usage of the new functionalities for potential users.

For vanishing molecules in solution, the VI method developed in chapter 5 yielded accuracies that were similar to the ones of established methods, suggesting that it does not represent the optimum in this context, yet. The underlying reasons will be investigated in chapter 6. Furthermore, the VI method will be applied to non-equilibrium applications, and its accuracy assessed.

2

Theory and Methods

This chapter will lay out the underlying theory and extend upon the concepts and methods from the introduction that a) will either be used in this thesis, b) the derivations in the following chapters are based on, or c) are alternatives to the linear interpolation scheme, which the VI method derived in this work will be compared to. Furthermore, MD simulations, which will be predominantly used to conduct sampling for atomistic systems, will be described.

2.1 Definitions

Extensive descriptions of the statistical physics background and context of free energies can be found in numerous textbooks, such as the ones by Landau et al. [160] and Nolting [161]. The essential definitions will be repeated here.

A canonical ensemble describes a system in contact with a thermal reservoir of constant temperature T, where energy can be freely exchanged. It is assumed that the reservoir is much larger than the system of interest such that the temperature change of the reservoir due to the energy exchange is negligible. This assumption is fulfilled for the cases of interest for this thesis, as, e.g., a single protein is much smaller than the surrounding cell or organism.

A classical Hamiltonian system with n particles is considered. Assuming thermal equilibrium with the heat bath, the probability p of a microstate, characterized by the $3n$ dimensional canonical positions $\mathbf{x}$ and conjugate momenta $\mathbf{p}$, is given by

$$p(\mathbf{x}, \mathbf{p}) = \frac{1}{h^3 Z} e^{-\beta H(\mathbf{x}, \mathbf{p})} \ , \tag{2.1}$$

where h is the Planck constant, $\beta = \frac{1}{k_B T}$ the reciprocal temperature, k_B the Boltzmann constant and $H(\mathbf{x}, \mathbf{p})$ the Hamiltonian. The exponential term is referred to as the Boltzmann factor. For indistinguishable particles, the prefactor reads as $\frac{1}{n! h^{3n} Z}$. The partition function

$$Z = \frac{1}{h^3} \int_{-\infty}^{\infty} \int_{-\infty}^{\infty} e^{-\beta H(\mathbf{x}, \mathbf{p})} \mathrm{dx}\, \mathrm{dp} \tag{2.2}$$

serves as a normalization constant in Eq. 2.1. The term $1/h^3$ cancels in all derivations in this thesis. For ease of notation, it is therefore omitted in the remaining parts of this work.

The free energy is defined via the partition function as

$$F = -\beta^{-1} \ln Z \ , \tag{2.3}$$

which, upon inserting Z, Eq. 2.2, yields Eq. 1.1 from the introduction.

Using macroscopic variables,

$$F = H - TS \ , \tag{2.4}$$

where H denotes the enthalpy and S the entropy. For an NVT ensemble (i.e., a system with fixed particle number, volume and temperature), F is referred to as the Helmholtz free energy. In this case, the enthalpy reduces to the internal energy $H = U$.

The Gibbs free energy corresponds to an NPT ensemble, and is generally denoted by G. Here, $H = U + PV$, where the additional term PV accounts for the energy corresponding to a change in volume. Equations 2.2 and 2.3 also hold for the Gibbs free energy, provided that the Hamiltonian $H_G(\mathbf{x}, \mathbf{p}) = H(\mathbf{x}, \mathbf{p}) + PV$ is defined such that it includes the correction term. As most biophysical processes and experiments are conducted at constant pressure and temperature, all further definitions will be based on the Gibbs free energy, but can be applied similarly to the Helmholtz free energy, too.

2.2 Equilibrium Sampling Approaches

Free Energy Perturbation

Free energy perturbation (FEP) refers to the methodology that rests on the Zwanzig formula [79]. As the derivations in chapter 3 are largely based on this formula, it's derivation and context will shortly be outlined below in this subsection. Importantly, note that in contrast to what the name "perturbation" suggests, the Zwanzig formula is exact, and as such its derivation does not contain any approximations. Instead, the name was based on the fact that, when it was first developed, the Zwanzig formula provided the starting point for a perturbation approach leading to alternative approximated variants thereof [79, 90]. However, for computational free energy calculations, in the large majority of cases the exact form is used, yet the name "perturbation" remains. Extensive descriptions on the background and applications of the following methodology can be found in the reviews by Chipot and Pohorille [1], Christ et al. [7] and Gapsys et al. [162] or in the best practice guides by Shirts et al. [163] and Mey et al. [164].

For two states A and B, with the corresponding partition functions Z_A and Z_B, respectively, the difference in the Gibbs free energy is, using Eq. 2.3,

$$\Delta G = -\beta^{-1} \ln \frac{Z_B}{Z_A} \ . \tag{2.5}$$

With the definition of the partition function, Eq. 2.2,

$$\Delta G = -\beta^{-1} \ln \frac{\int_{-\infty}^{\infty}\int_{-\infty}^{\infty} e^{-\beta H_B(\mathbf{x},\mathbf{p})}\, d\mathbf{x}\, d\mathbf{p}}{\int_{-\infty}^{\infty}\int_{-\infty}^{\infty} e^{-\beta H_A(\mathbf{x},\mathbf{p})}\, d\mathbf{x}\, d\mathbf{p}} \tag{2.6}$$

$$= -\beta^{-1} \ln \frac{\int_{-\infty}^{\infty}\int_{-\infty}^{\infty} e^{-\beta[H_B(\mathbf{x},\mathbf{p})-H_A(\mathbf{x},\mathbf{p})]} e^{-\beta H_A(\mathbf{x},\mathbf{p})}\, d\mathbf{x}\, d\mathbf{p}}{\int_{-\infty}^{\infty}\int_{-\infty}^{\infty} e^{-\beta H_A(\mathbf{x},\mathbf{p})}\, d\mathbf{x}\, d\mathbf{p}} \tag{2.7}$$

$$= -\beta^{-1} \ln \int_{-\infty}^{\infty}\int_{-\infty}^{\infty} e^{-\beta[H_B(\mathbf{x},\mathbf{p})-H_A(\mathbf{x},\mathbf{p})]} p_A(\mathbf{x},\mathbf{p})\, d\mathbf{x}\, d\mathbf{p} \tag{2.8}$$

$$= -\beta^{-1} \ln \left\langle e^{-\beta[H_B(\mathbf{x},\mathbf{p})-H_A(\mathbf{x},\mathbf{p})]} \right\rangle_A \ , \tag{2.9}$$

where $\langle ... \rangle_A$ denotes the ensemble average of A. Equation 2.9 is the Zwanzig formula

[79]. The ensemble average can either be calculated analytically, or by sampling in state A using an MC or MD based approach. In this thesis, the state that the ensemble average is calculated of (i.e. A in Eq. 2.9) is referred to as the reference state, whereas B is the target state.

In the following chapters, the Zwanzig formula, Eqs. 2.9, will be expressed only as a function of $H(\mathbf{x})$. The reason is based on the decomposition of $H(\mathbf{x}, \mathbf{p}) = T(\mathbf{p}) + V(\mathbf{x})$, where $T(\mathbf{p})$ denotes the kinetic energy contribution. As the kinetic term only depends on particle momenta, it can be integrated out analytically. As the individual masses do not change between A and B in all applications of the following chapters, these contributions cancel out, and will therefore be omitted in the above formalism. In cases where masses do change, the average kinetic energy is still the same in A and B, due to the equipartition theorem and coupling to the heat bath. However, as the temperature of the system within the heat bath is fluctuating, the kinetic energy contributions to A and B do not generally cancel out and, given the non-linearity of Eq. 2.9, still have to be considered. Therefore, this chapter uses the more general formulation.

Along similar lines, the expectation value of other observables $O(\mathbf{x}, \mathbf{p})$ of B can be calculated as

$$\langle O(\mathbf{x}, \mathbf{p}) \rangle_B = \frac{\left\langle O(\mathbf{x}, \mathbf{p})\, e^{-\beta[H_B(\mathbf{x}, \mathbf{p}) - H_A(\mathbf{x}, \mathbf{p})]} \right\rangle_A}{\left\langle e^{-\beta[H_B(\mathbf{x}, \mathbf{p}) - H_A(\mathbf{x}, \mathbf{p})]} \right\rangle_A}, \tag{2.10}$$

i.e., based on an ensemble average in A.

As outlined in the introduction, a sequence of N states, i.e., $N-2$ intermediates, will be used to improve sampling accuracy. In this case, the total difference is calculated as.

$$\Delta G = \sum_{s=1}^{N-1} \Delta G_{s, s+1} \tag{2.11}$$

$$= -\beta^{-1} \sum_{s=1}^{N-1} \ln \left\langle e^{-\beta[H_{s+1}(\mathbf{x}, \mathbf{p}) - H_s(\mathbf{x}, \mathbf{p})]} \right\rangle_s, \tag{2.12}$$

where $\Delta G_{s, s+1}$ denotes the free energy difference between states s and $s + 1$ with $s = 1$ denoting state A and $s = N$ denoting state B. Similarly, the difference between A and B of any other ensemble based observable can be calculated by using Eq. 2.10 in a step-wise approach. Whereas the terminology is sometimes used ambivalently, in the context of this thesis FEP refers to using the Zwanzig formula in multiple steps.

Paths and Sequences

This thesis distinguishes between alchemical paths and sequences. The first refers to a (continuous) transformation of the Hamiltonians along a path variable λ, such as the linear interpolation scheme, $H_\lambda(\mathbf{x}, \mathbf{p}) = (1 - \lambda)H_A(\mathbf{x}, \mathbf{p}) + \lambda H_B(\mathbf{x}, \mathbf{p})$, described in the introduction.

For FEP and other equilibrium techniques, sampling is conducted in a series of N states governed by the Hamiltonians $H_1(\mathbf{x}, \mathbf{p}), ..., H_N(\mathbf{x}, \mathbf{p})$. Such a series of states will be referred to as a sequence. In practice, these Hamiltonians are almost exclusively characterized by a set of λ points $\{\lambda_1, ..., \lambda_N\}$ (with $\lambda_1 = 0$ and $\lambda_N = 1$) along a path, and usually along the linearly interpolated one. The choice of the intermediate λ points has been the subject to a number of optimization approaches [6, 165].

This thesis distinguishes between path and sequence, as for the latter one, any definition of intermediate Hamiltonians can be used without assuming a path *a priori*. In chapter 3, it will be shown that, in fact, the sequence yielding the optimal MSE — the VI sequence — is such a case and, in its exact form, does not require any choice of a path variable.

Bennett Acceptance Ratio Method

In the limit of an infinite number of sample points, the Zwanzig formula, Eq. 2.9, yields the exact result no matter if using A or B as the reference state that sampling is conducted in. However, for finite sampling, the convergence properties of both variants generally differ. Except for a number of special cases, the free energy estimates improve if the information from samples in both states are used. The main estimator for this purpose is the Bennett Acceptance Ratio (BAR) method. It will be extensively used throughout this thesis and chapter 3 will provide an alternative derivation and generalization thereof. BAR will therefore be briefly described below.

Bennett started by expanding the definition of the difference in free energy

$$\Delta G = -\beta^{-1} \ln \frac{Z_B}{Z_A} \tag{2.13}$$

$$= -\beta^{-1} \ln \frac{Z_B \int_{-\infty}^{\infty} \int_{-\infty}^{\infty} w[H_A(\mathbf{x},\mathbf{p}), H_B(\mathbf{x},\mathbf{p})] e^{-H_A(\mathbf{x},\mathbf{p}) - H_B(\mathbf{x},\mathbf{p})} \, d\mathbf{x} \, d\mathbf{p}}{Z_A \int_{-\infty}^{\infty} \int_{-\infty}^{\infty} w[H_A(\mathbf{x},\mathbf{p}), H_B(\mathbf{x},\mathbf{p})] e^{-H_A(\mathbf{x},\mathbf{p}) - H_B(\mathbf{x},\mathbf{p})} \, d\mathbf{x} \, d\mathbf{p}} \tag{2.14}$$

$$= \beta^{-1} \ln \frac{\left\langle w[H_A(\mathbf{x},\mathbf{p}), H_B(\mathbf{x},\mathbf{p})] e^{-H_A(\mathbf{x},\mathbf{p})} \right\rangle_B}{\left\langle w[H_A(\mathbf{x},\mathbf{p}), H_B(\mathbf{x},\mathbf{p})] e^{-H_B(\mathbf{x},\mathbf{p})} \right\rangle_A} \tag{2.15}$$

by an arbitrary weighting function w. Optimizing w such that the estimates in the free energy difference yield the smallest variance leads to the BAR method

$$\Delta G = \beta^{-1} \ln \frac{\left\langle f[\beta(H_A(\mathbf{x},\mathbf{p}) - H_B(\mathbf{x},\mathbf{p}) + C)] \right\rangle_B}{\left\langle f[\beta(H_B(\mathbf{x},\mathbf{p}) - H_A(\mathbf{x},\mathbf{p}) - C)] \right\rangle_A} + C \;, \tag{2.16}$$

where $f(x) = 1/(1 + \exp(x))$ is the Fermi function, and the variance of the free energy estimate minimal if

$$C = -\beta^{-1} \ln \left(\frac{Z_B n_A}{Z_A n_B} \right) \tag{2.17}$$

$$= \Delta G - \beta^{-1} \ln \left(\frac{n_A}{n_B} \right) \;. \tag{2.18}$$

The numbers n_A and n_B denote the number of independent sample points in states A and B, respectively. Due to the dependence of C on ΔG, equation 2.16 is an implicit problem. It is solved iteratively by using an initial guess for C, and then updating C by using Eqs. 2.16 and 2.18 until convergence. Equivalently, the equation

$$\sum_{i=1}^{n_A} f(\beta[H_A(\mathbf{x}_i, \mathbf{p}_i) - H_B(\mathbf{x}_i, \mathbf{p}_i) + C])$$
$$= \sum_{j=1}^{n_B} f(\beta[H_A(\mathbf{x}_j, \mathbf{p}_j) - H_B(\mathbf{x}_j, \mathbf{p}_j) - C]) \;, \tag{2.19}$$

where i and j label the samples obtained from state A and B, respectively, can be solved numerically for C. The difference ΔG is then subsequently calculated through Eq. 2.18.

When using $N > 2$ states as in Eq. 2.12, given a sample point $\mathbf{x}_t, \mathbf{p}_t$ from state s, it is not only possible to calculate the Hamiltonians $H_s(\mathbf{x}_t, \mathbf{p}_t)$ and $H_{s+1}(\mathbf{x}_t, \mathbf{p}_t)$ of the states s and $s + 1$, but instead of all states $s = 1, ..., N$. The estimator — the multistate Bennett Acceptance Ratio (MBAR) [166] method — uses this information and was developed based on a related extended bridge sampling technique [167]. The free energies $\{G_s\}$ of all states s,

$$G_s = -\ln \sum_{t=1}^{N} \sum_{i=1}^{n_t} \frac{e^{-\beta H_s(\mathbf{x}_i^t, \mathbf{p}_i^t)}}{\sum_{k=1}^{N} n_k\, e^{G_k - \beta H_k(\mathbf{x}_i^t, \mathbf{p}_i^t)}}, \tag{2.20}$$

are calculated with respect to an unkown constant, where s, t and k are labels for states, and $\mathbf{x}_i^t$ and $\mathbf{p}_i^t$ denote the ith positions and momenta, respectively, from state t. The integer numbers n_k and n_t denote the total number of sample points from the state indicated by the subscript. Once all $\{G_s\}$ have been determined, the differences are obtained by, e.g., subtracting G_1 from all $\{G_s\}$, yielding the desired free energy difference between the end states through $G_N - G_1$. Again, the set of equations 2.20 is an implicit problem, and is solved iteratively.

Thermodynamic Integration

The most prominent equilibrium method next to FEP is Thermodynamic Integration (TI). It provides the basis of the minimum variance path (MVP) that will be described in the next section and used as a comparison to the VI sequence in chapter 3.

If the path is differentiable with respect to λ between 0 and 1, then

$$\frac{\partial G}{\partial \lambda} = -\beta^{-1} \frac{\partial}{\partial \lambda} \ln \int_{-\infty}^{\infty} \int_{-\infty}^{\infty} e^{-\beta H(\mathbf{x}, \mathbf{p}, \lambda)} d\mathbf{x}\, d\mathbf{p} \tag{2.21}$$

$$= \frac{\int_{-\infty}^{\infty} \int_{-\infty}^{\infty} e^{-\beta H(\mathbf{x}, \mathbf{p}, \lambda)} \dfrac{\partial H(\mathbf{x}, \mathbf{p}, \lambda)}{\partial \lambda} d\mathbf{x}\, d\mathbf{p}}{\int_{-\infty}^{\infty} \int_{-\infty}^{\infty} e^{-\beta H(\mathbf{x}, \mathbf{p}, \lambda)} d\mathbf{x}\, d\mathbf{p}} \tag{2.22}$$

$$= \left\langle \frac{\partial H(\mathbf{x}, \lambda)}{\partial \lambda} \right\rangle_{\lambda} \tag{2.23}$$

$$\Rightarrow \Delta G = \int_{0}^{1} \left\langle \frac{\partial H(\mathbf{x}, \mathbf{p}, \lambda)}{\partial \lambda} \right\rangle_{\lambda} d\lambda \ , \tag{2.24}$$

where Eq. 2.24 is referred to as Thermodynamic Integration (TI) [90]. Note that TI is derived based on an equilibrium average for each λ, while integrating continuously along λ. In practice, sampling is therefore also conducted at discrete steps in equilibrium, while Eq. 2.24 is calculated via numerical integration using, e.g., the Simpson or the trapezoid rule.

2.3 Soft-core Interactions and other Choices of Intermediate States

All methods mentioned above contain evaluations of a sample that was generated in a reference state s with a Hamiltonian of a different target state, e.g., $s+1$. Furthermore, for the common linear interpolation scheme, to determine $H_{s+1}(\mathbf{x}, \mathbf{p})$, both $H_A(\mathbf{x}, \mathbf{p})$ and $H_B(\mathbf{x}, \mathbf{p})$ need to be calculated. In the following it is assumed that the system contains vanishing particles, as illustrated in Fig. 2.1 on the example of calculating solvation free energies. Therefore, particles coupled through regular LJ interactions in B are represented as decoupled "ghost" particles in A with zero interaction energies.

The linearly interpolated intermediates of such a vanishing LJ particle are shown in Fig. 2.2(a). When sampling is conducted in A, represented by the flat blue line, these "ghost" particles will overlap with other ones. In these cases, complications will arise if such a configuration from A would be evaluated at $H_B(\mathbf{x}, \mathbf{p})$ (red line), due to the divergence at $r = 0$ of the LJ term that accounts for the Pauli repulsion. Furthermore, when decoupling a charged atom or particle, these complications would be enhanced in intermediate states closer to A in the sequence, if

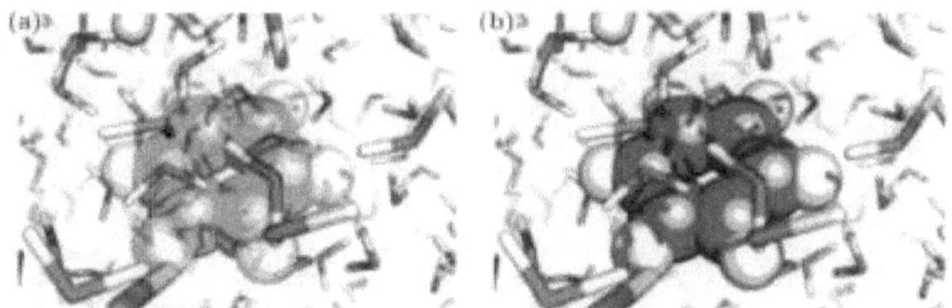

Figure 2.1: End states for calculating the solvation free energy difference through decoupling LJ interactions. (a) Decoupled state, representing the state where the solute has been removed from water. The solute does not interact with its surrounding, therefore it overlaps with some water molecules. (b) Coupled state: The solute interacts regularly with water. Image courtesy Dr. Vytautas Gapsys.

a remaining attractive electrostatic interaction was strong enough to overcome the weak Pauli repulsion down to small center–center separations that could otherwise not occur.

To avoid the problem for electrostatic interactions, these can be decoupled separately in a first step while using full LJ interactions, which are decoupled in the second step [168, 169]. More importantly, soft-core potentials [91, 92] can be used by replacing the divergence at $r = 0$ with a finite maximum, high enough such that it will never be visited for state B, but becoming low enough such that it will be visited for intermediate states. To this aim, a λ dependence of the end states is introduced, such that the intermediate Hamiltonians are calculated as

$$H_\lambda(\mathbf{x}, \mathbf{p}) = (1 - \lambda) H_A(\mathbf{x}, \mathbf{p}, \lambda) + \lambda H_B(\mathbf{x}, \mathbf{p}, \lambda) \ . \tag{2.25}$$

In the LJ potential

$$V^{LJ}(r_{ij}) = 4\epsilon_{ij} \left[\left(\frac{\sigma_{ij}}{r_{ij}} \right)^{12} - \left(\frac{\sigma_{ij}}{r_{ij}} \right)^{6} \right] \ , \tag{2.26}$$

where σ_{ij} and ϵ_{ij} denote the LJ parameter, the distance r_{ij} between two atoms i

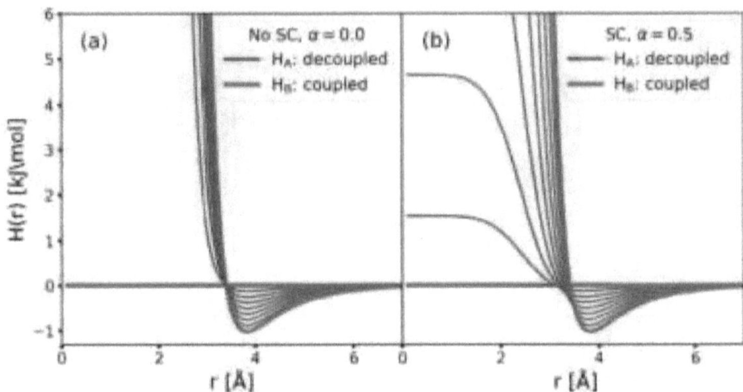

Figure 2.2: Intermediate interaction energies at distance r of a vanishing particle without and with soft-core (SC) treatment. In state B (red line), it is fully interacting through a LJ potential. In contrast, in A (blue line), the interaction energy is zero at all r. The lines in between indicate the intermediate Hamiltonians obtained through (a) the linear interpolation scheme, Eq. 1.3, and (b) the soft-core variant thereof, i.e., Eq. 2.25.

and j is replaced by two altered radii

$$r_A(r_{ij}, \lambda) = \left(\alpha \sigma_A^c \lambda^b + r_{ij}^c\right)^{\frac{1}{c}},$$
(2.27)

$$r_B(r_{ij}, \lambda) = \left(\alpha \sigma_B^c (1 - \lambda)^b + r_{ij}^c\right)^{\frac{1}{c}},$$
(2.28)

for each end state. Here, α, b and c are dimensionless soft-core parameters, and σ_A and σ_B the LJ parameters in the end states A and B, respectively. As can be seen, both r_A and r_B are finite at $r_{ij} = 0$, for all intermediate λ points, thereby avoiding divergences. The resulting Hamiltonians for a sequence of equally spaced λ values are shown in Fig. 2.2(b). The correct end states without alteration are reproduced at $\lambda = 0$ for A, and $\lambda = 1$ for B. The intermediate form without soft-core is obtained for $\alpha = 0$.

In the form by Steinbrecher et al. [93], Eq. 2.25 was further generalized by replacing the prefactors with $(1 - \lambda)^a$ and λ^a, where a is another soft-core parameter. However, to date $a = 1$, $\alpha = 0.5$, $b = 1$ or 2, and $c = 6$ are considered optimal [164, 170, 171] and also used in this thesis, such that the generalization by a did not reveal any significant improvements, yet.

For $\sigma_A = 0$ and $\sigma_B = \sigma_{ij}$, as well as $\epsilon_A = 0$ and $\epsilon_B = \epsilon_{ij}$, as in the example above, the LJ interactions in the intermediate state are calculated as

$$V_\lambda^{LJ}(r_{ij}) = (1 - \lambda)V^{LJ}(r_A(r_{ij})) + \lambda V^{LJ}(r_B(r_{ij})) \tag{2.29}$$

$$= 4\,\lambda\,\epsilon_{ij} \left(\left[\alpha(1-\lambda)^b + \left(\frac{r_{ij}}{\sigma_{ij}}\right)^c\right]^{-\frac{12}{c}} - \left[\alpha(1-\lambda)^b + \left(\frac{r_{ij}}{\sigma_{ij}}\right)^c\right]^{-\frac{6}{c}} \right) . \tag{2.30}$$

This form is the one most commonly found in the literature, whereas Eqs. 2.27 and 2.28 present the more general one.

If the decoupling of LJ and Coulomb interactions should be conducted simultaneously, then the soft-core approach can be extended to Coulomb interactions as well. In this case, these are also calculated based on the modified distances $r_A(r_{ij}, \lambda)$ and $r_B(r_{ij}, \lambda)$. Assuming again that A represents the decoupled state, where the charges do not to interact with their surrounding, the interaction is calculated along λ as

$$V_\lambda^{coul}(r_{ij}) = \frac{q_i\,q_j}{4\pi\epsilon_0\epsilon_r \left[\alpha(1-\lambda)^b + r_{ij}^c\right]^{\frac{1}{c}}} , \tag{2.31}$$

where q_i and q_j denote the charges in the coupled state, and ϵ_0 and ϵ_r the vacuum and relative permittivity, respectively.

Soft-core potentials will be used to calculate solvation free energies in chapter 3 and 5. In addition, an approach similar in principle will be devised and implemented in chapter 5.

Minimum Variance Path

The soft-core variant, Eq. 2.25 offers more flexibility in devising an alchemical path through regulating the λ dependence of the end states than the regular linear interpolation path, Eq. 1.3. However, it still represents only a very special case among all possible ones. The most relevant one for this work — the Minimum Variance Path (MVP) — was derived by Blondel [104] and will emerge as a limiting case of the VI sequence derived in chapter 3. States from the MVP will be used for a comparison in accuracy. Its derivation is therefore shortly outlined below.

Blondel aimed at minimizing the variance of the free energy estimates,

$$\int_0^1 \left\langle \left(\frac{\partial H_\lambda(\mathbf{x},\mathbf{p})}{\partial \lambda} \right)^2 - \left\langle \frac{\partial H_\lambda(\mathbf{x},\mathbf{p})}{\partial \lambda} \right\rangle_\lambda^2 \right\rangle_\lambda \mathrm{d}\lambda \tag{2.32}$$

$$= \int_0^1 \left\langle \left(\frac{\partial H_\lambda(\mathbf{x},\mathbf{p})}{\partial \lambda} \right)^2 \right\rangle_\lambda \mathrm{d}\lambda - \int_0^1 \left\langle \frac{\partial H_\lambda(\mathbf{x},\mathbf{p})}{\partial \lambda} \right\rangle_\lambda^2 \mathrm{d}\lambda \tag{2.33}$$

for TI. In the next step, he used the translation invariance: At each λ point, a bias function $B(\lambda)$, which is constant across configuration space, can be added to the Hamiltonian

$$H'_\lambda(\mathbf{x},\mathbf{p}) = H_\lambda(\mathbf{x},\mathbf{p}) + B(\lambda) \ . \tag{2.34}$$

The derivatives of the bias function will cancel out in Eq. 2.32 but can be designed such that the second term for the altered Hamiltonian,

$$\left\langle \frac{\partial H'_\lambda(\mathbf{x},\mathbf{p})}{\partial \lambda} \right\rangle_\lambda = 0 \ , \tag{2.35}$$

thereby simplifying the derivation. Next, considering the first term in Eq. 2.33,

$$\int_0^1 \left\langle \left(\frac{\partial H'_\lambda(\mathbf{x},\mathbf{p})}{\partial \lambda} \right)^2 \right\rangle_\lambda \mathrm{d}\lambda \tag{2.36}$$

$$= \int_0^1 \int_{-\infty}^{\infty} \int_{-\infty}^{\infty} \left(\frac{\partial H'_\lambda(\mathbf{x},\mathbf{p})}{\partial \lambda} \right)^2 e^{-\beta H'_\lambda(\mathbf{x},\mathbf{p})} \mathrm{d}\mathbf{x}\,\mathrm{d}\mathbf{p}\,\mathrm{d}\lambda \tag{2.37}$$

$$= \int_{-\infty}^{\infty} \int_{-\infty}^{\infty} \int_0^1 \left(\frac{\partial H'_\lambda(\mathbf{x},\mathbf{p})}{\partial \lambda} e^{-\beta \frac{H'_\lambda(\mathbf{x},\mathbf{p})}{2}} \right)^2 \mathrm{d}\lambda\,\mathrm{d}\mathbf{x}\,\mathrm{d}\mathbf{p} \ , \tag{2.38}$$

where, in the last step, Blondel assumed that the integrals converge to a finite value and can therefore be exchanged. Furthermore, for any point,

$$\int_0^1 \frac{\partial H'_\lambda(\mathbf{x},\mathbf{p})}{\partial \lambda} e^{-\beta \frac{H'_\lambda(\mathbf{x},\mathbf{p})}{2}} \mathrm{d}\lambda = -2\beta^{-1} \left(e^{-\frac{\beta}{2}H_A(\mathbf{x},\mathbf{p})} + e^{-\frac{\beta}{2}H_B(\mathbf{x},\mathbf{p})} \right) \tag{2.39}$$

$$:= C \ . \tag{2.40}$$

Hence, in the next step, the inner integral in Eq. 2.38 of the form

$$\int_0^1 f(\lambda)^2 \mathrm{d}\lambda \tag{2.41}$$

can be minimized by setting

$$f(\lambda) = C \;\Rightarrow\; \int_0^1 f(\lambda)\mathrm{d}\lambda = C \tag{2.42}$$

as for any other function

$$\int_0^1 g(\lambda)^2\mathrm{d}\lambda = C^2 + \int_0^1 [g(\lambda) - C]^2\mathrm{d}\lambda > C^2 \;. \tag{2.43}$$

Blondel now obtained $H_\lambda(\mathbf{x}, \mathbf{p})$ by integrating Eq. 2.39 from 0 to λ,

$$-2\beta^{-1}\left(e^{-\frac{\beta}{2}H'_\lambda(\mathbf{x},\mathbf{p})} - e^{-\frac{\beta}{2}H_A(\mathbf{x},\mathbf{p})}\right) = -\lambda 2\beta^{-1}\left(e^{-\frac{\beta}{2}H_B(\mathbf{x},\mathbf{p})} - e^{-\frac{\beta}{2}H_A(\mathbf{x},\mathbf{p})}\right) \tag{2.44}$$

yielding the Hamiltonian

$$H_\lambda(\mathbf{x}, \mathbf{p}) = -2\beta^{-1}\ln\left((1-\lambda)e^{-\frac{\beta}{2}H_A(\mathbf{x},\mathbf{p})} + \lambda e^{-\frac{\beta}{2}H_B(\mathbf{x},\mathbf{p})}\right) + \lambda\Delta\hat{G}^2 \;, \tag{2.45}$$

where the last term was added to transform from $H'_\lambda(\mathbf{x}, \mathbf{p})$ to $H(\mathbf{x}, \mathbf{p})$ again, which was required for Eq. 2.35. Therefore, the minimum variance path (MVP) is optimal if $\Delta\hat{G} \approx \Delta G$.

Later, Pham and Shirts [105] also derived this path by adapting a result from Gelman and Meng [78]. Furthermore, they established that Eq. 2.45 yields the optimal variance for high configuration space overlap between the end states, whereas in the low overlap case, the prefactors $\cos(\lambda\frac{\pi}{2})$ and $\sin(\lambda\frac{\pi}{2})$ are preferred instead of $(1-\lambda)$ and λ.

Note that these paths differ only for continuous and equal sampling along λ. However, when using discrete states, as general practice for TI, any pair of prefactors ζ_1 and ζ_2 can be rescaled by the factor $r = 1/(\zeta_1 + \zeta_2)$ such that $r\zeta_1 + r\zeta_2 = 1$. The term $\ln(r)$ will appear as an additive constant in Eq. 2.45, and therefore will drop out for all intermediates when calculating the step-wise free energy difference between the end states. Therefore, redefining $\lambda := r\zeta_2 \Rightarrow (1-\lambda) = r\zeta_1$ leads to the original form of Eq. 2.45. As such, for discrete states, any set of λ steps along the path designed for the low overlap case yields, on average, equivalent total estimates as another set of steps in the high overlap variant.

Enveloping Distribution Sampling

Other alternatives to Eq. 2.25 are the One-Step Perturbation (OSP) [97–99] and the Enveloping Distribution Sampling (EDS) [100, 101] class of methods. These are designed to calculate free energy differences from a single simulation by constructing a reference potential that "envelopes", i.e., combines all configuration space regions relevant to one or more end states. For example, in one application, OSP [99] was applied to compare different rigid ligands to an estrogen receptor. The reference state was designed by using a "ghost" particle at the positions were the atoms differ between the ligands. For EDS, the configuration space density of the reference state is constructed by summing over the ones from the N_i end states

$$e^{-\beta s H_{ref}(\mathbf{x},\mathbf{p})} = \sum_{i=1}^{N_i} e^{-\beta s (H_i(\mathbf{x},\mathbf{p}) - E_i)} , \qquad (2.46)$$

where s and E_i, for $i = 1, ..., N_i$, are adjustable parameters. The Zwanzig formula [79], Eq. 2.9, is used to calculate the free energy difference between the reference state and each end state.

If the configuration space regions of all end states were disjunct, equal sampling would, assuming uncorrelated sample points, be conducted in these regions if $E_i = G_i$. However, in other cases, the optimal E_i may differ for more than two end states. Furthermore, the parameter s can be adjusted to avoid barriers and typically ranges between 0.001 and 1 [101]. For small s, such that $\beta s H_i(\mathbf{x},\mathbf{p}) << 1$, a series expansion of the exponentials and the logarithm in Eq. 2.46 yields the linear interpolation

$$H_{ref}(\mathbf{x},\mathbf{p}) = \sum_{i=1}^{N_i} H_i(\mathbf{x},\mathbf{p}) + \text{const} . \qquad (2.47)$$

Therefore, s switches between the interpolation of the Hamiltonians (small s) and the configuration space densities ($s = 1$). A very recent variant, λ-EDS [103], which was constructed also based on our work in chapter 3, now also calculates the free energy difference between only two end states, but using several intermediate ones instead.

The usage of the s factor will also be combined with the VI sequence and be made accessible through the GROMACS implementation described in chapter 5.

2.4 Non-Equilibrium Approaches

The second class of methods that employ the concept of alchemical paths, as described in the introduction, are non-equilibrium techniques. Here, transitions from A to B are simulated by increasing λ at every step; the system is therefore out of equilibrium. In chapter 6, an approximated path deduced from the VI sequence will be tested with non-equilibrium methods.

Small perturbations away from equilibrium, such as for very slow transitions, can be described using linear response theory [172–176]. More generally, Jarzynski [84] derived the identity

$$e^{-\beta\Delta G} = \left\langle e^{-\beta W_{A\to B}} \right\rangle ,$$
(2.48)

where the brackets indicate an ensemble average over several transitions starting from A and ending at B. The work values are obtained similarly to TI by integration along λ

$$W = \int_0^1 \left\langle \frac{\partial H(\mathbf{x}, \lambda)}{\partial \lambda} \right\rangle_\lambda d\lambda .$$
(2.49)

The starting configurations need to represent the equilibrium ensemble. Fascinatingly, Eq. 2.48 holds arbitrarily far from equilibrium, which, in addition to the theory, has been demonstrated by experiments through reversible and irreversible pulling of an RNA molecule [177] for the first time.

To utilize the information from both forward and backward transitions, where $P_f(W)$ and $P_r(W)$ denote corresponding work distributions, the Crooks Fluctuation Theorem (CFT) [178, 179] (also valid for NPT ensembles [180])

$$\frac{P_f(W)}{P_r(-W)} = e^{\beta(W-\Delta G)}$$
(2.50)

provides the basis of several methods [181, 182]. For example, the Crooks Gaussian Intersection (CGI) estimator approximates $P_f(W)$ and $P_r(W)$ by two Gaussian functions and estimates the free energy difference through their intersection.

Furthermore, the BAR estimator was also adapted for non-equilibrium simulations [183], where the free energy is determined through numerical solution of

$$\sum_{i=1}^{n_f} \frac{1}{1 + \exp\left(\beta(W_i - C)\right)} = \sum_{j=1}^{n_r} \frac{1}{1 + \exp\left(-\beta(W_j - C)\right)} \, , \tag{2.51}$$

for C, which, similar to Eq. 2.18, relates to the free energy as

$$\Delta G = C + \beta^{-1} \ln\left(\frac{n_f}{n_r}\right) \, . \tag{2.52}$$

Here, n_f and n_r denote the numbers of transitions in forward and reverse directions, respectively. Furthermore, it was shown that the BAR estimator yields the free energy with the maximum likelihood, given the observed work values [183].

Both the BAR and CGI estimator will be used in chapter 6 to calculate free energy differences based on non-equilibrium trajectories.

Overlap Sampling

In an extensive series of publications [184–189], the group of Kofke and coworkers analyzed the scenario in which the relevant regions in configuration space of, e.g., state B, represent only a small subset, and are completely embedded within the ones of state A. Their example in this case consists of a vanishing particle, which in the decoupled state A has a much higher entropy as in state B, where its position is restricted by its neighboring particles.

Through a variety of theoretical considerations and simulations, they established that for one-sided simulations, the systematic error, i.e., the bias of the free energy estimates, is lower when starting from the state of higher, and targeting the state with lower entropy. For a vanishing particle, the trajectories should therefore be started from the decoupled state, in other words, insertion is preferable to deletion. In this special case, conducting only one-sided simulations has even a smaller bias than when trajectories from both sides are used with the approaches outlined in the beginning of this section.

Next, they considered the case in which the configuration space volumes of A and B are more similar, and these two overlap in a region O. Then, by definition, O is a subset region of both A and B. Therefore, when introducing O as an intermediate state and calculating the free energy difference through $A \rightarrow O$ and $B \rightarrow O$, where the arrow indicates a set of non-equilibrium transitions starting from the first and ending at the second state, then in both cases, the transitions start at the state of higher entropy and end at the one lower entropy. Using the use of the Jarzynski identity [84], Eq. 2.48,

$$e^{-\beta \Delta G} = \frac{\left\langle e^{-\beta W_{A \to O}} \right\rangle_A}{\left\langle e^{-\beta W_{A \to O}} \right\rangle_B} \ . \tag{2.53}$$

This approach is referred to as the Overlap Sampling (OS) method [186]. In analogy, the principle can be applied to FEP using the Zwanzig formula

$$e^{-\beta \Delta G} = \frac{\left\langle e^{-\beta [H_O(\mathbf{x},\mathbf{p}) - H_A(\mathbf{x},\mathbf{p})]} \right\rangle_A}{\left\langle e^{-\beta [H_O(\mathbf{x},\mathbf{p}) - H_B(\mathbf{x},\mathbf{p})]} \right\rangle_B} \ . \tag{2.54}$$

Note that in this case, sampling is conducted only in A and B. Therefore, O represents a virtual intermediate, i.e., a state that no sampling is conducted in, but which is used as a target state.

They suggested a simple form for O that uses the average of the end states $H_O(\mathbf{x},\mathbf{p}) = (H_A(\mathbf{x},\mathbf{p}) + H_B(\mathbf{x},\mathbf{p}))/2$, which already considerably improved the bias compared to calculating the estimate without the virtual intermediate. Similar to BAR, they extended the definition of O through a weighting function w

$$H_O(\mathbf{x},\mathbf{p}) = -\beta^{-1} \ln w[H_B(\mathbf{x},\mathbf{p}) - H_B(\mathbf{x},\mathbf{p})] + \frac{1}{2} [H_A(\mathbf{x},\mathbf{p}) + H_B(\mathbf{x},\mathbf{p})] \ , \tag{2.55}$$

where the two approaches are equivalent if

$$w(\Delta H(\mathbf{x},\mathbf{p})) = \frac{1}{\cosh\left[\dfrac{\beta}{2}(\Delta H(\mathbf{x},\mathbf{p}) - C)\right]} \ , \tag{2.56}$$

which is the Gaussian hyperbolic secant function, and where C corresponds to the definition of Bennett [80], Eq. 2.18.

Virtual intermediate states and their connections to estimators will be employed to construct the setup for free energy calculations underlying the VI and cVI sequences in chapter 3 and 4, respectively.

2.5   Molecular Dynamics Simulations

An important method for sampling used in this work are MD simulations. These model at a high level of detail — describing both the biomolecules and the solvent on an atomistic basis — while still being based on a classical many particle description of the system. It is important to distinguish that the underlying phenomena, e.g., the stretching of a bond, are non-classical.

MD simulations can, with current computer technology and algorithms, simulate time scales on the order of 50 to 100 ns per day for a two million atom system (on the example of the GROMACS 2018 software package with 8 standard CPUs and a GPU [68]). While simulations going beyond these limits have been conducted, these usually require substantial computational efforts. For example, first simulations above one millisecond were already conducted in 2011 [190] for several protein folding processes (system size of the order of ten thousand atoms) on the ANTON supercomputer [191], whose hardware was specifically optimized for the properties of MD. As for scale: recently (2019), a proof of concept run above one billion atoms of a gene locus was conducted [192]; however, progressing only at one nanosecond per day on 130,000 processor cores.

A large number of MD software packages has been developed such as NAMD [193, 194], AMBER [195], GROMOS [196], LAMMPS [197, 198] or OpenMM [199]. The one used and extended in this thesis is the GROMACS [157–159] software package. Its main advantage is its wide spread usage and its high speed compared to other packages. In contrast, alternatives, such as OpenMM, offer a higher degree of developer and user friendliness instead, where, e.g., integrators can be easily modified through a python interface. Therefore, a disadvantage of GROMACS is that such a change would, at least for the GROMACS 2019 and 2020 version, require more substantial changes in the source code than for, e.g., OpenMM.

Foundations and Approximations

Extensive descriptions about the fundamentals of MD simulations can be found in the recent book by Leimkuhler and Matthews [200], or the slightly older ones by Rapaport [201] or Frenkel and Smit [202]. A brief summary of the underlying principles and approximations of MD, and in extension, of the abilities and limitations of MD based free energy calculations, will be provided below.

As a first approximation, gravitational interactions are neglected, as these are, for example, roughly an order of 40 magnitudes weaker than electrostatic ones. Secondly, relativistic effects are considered negligible, due to the small masses and small velocities compared to the speed of light.

With these two approximations, systems at atomic length scales are described by the time-dependent Schrödinger equation [203]

$$i\hbar \frac{\partial}{\partial t}\psi(\mathbf{r}_e;\mathbf{r}_n) = \hat{H}\psi(\mathbf{r}_e;\mathbf{r}_n),\tag{2.57}$$

where $\hbar = \frac{h}{2\pi}$ denotes the reduced Planck constant and $\hat{H}$ the Hamiltonian operator. The wave-function $\psi(\mathbf{r}_e;\mathbf{r}_n)$ that describes the system depends on the positions of both the nuclei and the electrons $\mathbf{r}_n$, and $\mathbf{r}_e$, respectively.

Next, the fact that electrons are around 2000 times lighter than a nucleon and, therefore, move much faster, is considered. For this reason, the Born-Oppenheimer approximation [204] assumes that the electronic wave function instantaneously follows the motion of the nuclei, allowing to separate the wave function

$$\psi(\mathbf{r}_e;\mathbf{r}_n) = \psi_n(\mathbf{r}_n)\,\psi_e(\mathbf{r}_e;\mathbf{r}_n)\tag{2.58}$$

into the nuclear wave-function $\psi_n(\mathbf{r}_n)$ and the electronic wave-function $\psi_e(\mathbf{r}_e;\mathbf{r}_n)$, where the latter only depends on $\mathbf{r}_n$ parametrically. The decoupling of the time scales allows us to determine ψ_e with the time-independent Schrödinger equation

$$\hat{H}_e\psi_e(\mathbf{r}_e;\mathbf{r}_n) = E_e(\mathbf{r}_n)\psi_e(\mathbf{r}_e;\mathbf{r}_n)\,.\tag{2.59}$$

where $E_e(\mathbf{r}_n)$ denotes the ground state energy. Based on the Born-Oppenheimer approximation, the position of the nuclei is described by the time-dependent Schrödinger equation

$$\left(\hat{T}_n + E_e(\mathbf{r}_n)\right)\psi_n(\mathbf{r}_n) = i\hbar\frac{\partial}{\partial t}\psi(\mathbf{r}_n)\,.\tag{2.60}$$

where $\hat{T}_n$ denotes the kinetic energy operator of the nuclear motion.

Note that the Born-Oppenheimer approximation rests on the assumption that all electrons remain in the same state the entire time. This assumption is valid for temperatures around 300 K, where most of the biological systems operate. Here, given approximately 0.025 eV thermal energy per electron, the electrons are in the ground state, and as excitation energies generally exceed 1 eV (e.g., 10.2 eV for hydrogen), spontaneous excitation of electrons is unlikely. However, for the same reason, processes such as photon absorption cannot be addresses with standard MD.

Equation 2.60 describes the evolution of the entire nuclear wave-function. However, for applications targeted by MD simulations, it is sufficient to calculate the expectation value of the atom positions, now represented by the position of their nuclei under the above assumptions. These expectation values evolve according to the Ehrenfest theorem [205]

$$\frac{d\langle \mathbf{r}_n \rangle}{dt} = \langle \mathbf{v}_n \rangle \tag{2.61}$$

and

$$\frac{d\langle \mathbf{v}_n \rangle}{dr} = -\frac{\nabla V(\mathbf{r}_n)}{m} \, , \tag{2.62}$$

where $\mathbf{v}_n$ denotes the velocity and m the mass of the nuclei in a potential $V(\mathbf{r}_n)$. Given the similarity of Eqs. 2.61 and 2.62 to Newton's (classical) equations of motion, the evolution of the expectation values of an atom position and velocity can be considered as the ones of a point particle.

Integration of Equations of Motion

Newtons second equation of motion,

$$m_i \frac{\partial^2 \mathbf{r}_i}{\partial t^2} = \mathbf{F}_i \, , \tag{2.63}$$

where $i = 1, ..., n$, is solved numerically for all n atoms with mass m_i and force $\mathbf{F}_i = \nabla \mathbf{r}_i$ acting on atom i. The most widely used integrator for MD simulations is the leap-frog algorithm, a second-order integrator and equivalent to the velocity Verlet method. The positions of the atoms are propagated as

$$\mathbf{r}_i(t + \Delta t) = \mathbf{r}(t) + \mathbf{v}_i \left(t + \frac{\Delta t}{2} \right) \Delta t \tag{2.64}$$

and the velocities as

$$\mathbf{v}_i \left(t + \frac{\Delta t}{2} \right) = \left(t - \frac{\Delta t}{2} \right) + \frac{\mathbf{F}_i(t)}{m_i} \Delta t \, . \tag{2.65}$$

In this work, a time step of 2 fs is used in MD simulations. This value is typically used to keep the integration error substantially below errors caused by other approximations of MD simulations, such as the ones from force fields.

Force Fields

The interaction potential and the gradient thereof to calculate the force in Eq. 2.65 are determined by quantum mechanical effects. Calculating these is completely impractical for molecules consisting of more than a few hundred atoms, and therefore also for most of the applications of free energy calculations listed in the introduction. Therefore, force fields have been designed, that are an approximation to the ground state energy of the electrons. Their parameters are determined either from *ab initio* calculations [49, 50] for the interaction between individual atoms or small chemical groups, or through adjusting the parameters such that experimental observables [47, 48] of small model systems are reproduced (or, of course, a mixture of both approaches).

Conventional force fields divide the interaction potentials into bonded and non-bonded interactions. The former depend on bond lengths, Eq. (2.67), bond angles (2.68), dihedral angles (2.69) and improper dihedrals (2.70) that restrict out-of-plane motions to, e.g., keep aromatic rings planar. The non-bonded interactions consist of electrostatic (2.71), van-der-Waals interactions and Pauli repulsion, with the latter two described by a Lennard-Jones potential (2.72), which has already been described in section 2.2 for soft-core potentials. In total,

$$V(\mathbf{r}) = V_{\text{bonded}}(\mathbf{r}) + V_{\text{nonbonded}}(\mathbf{r}) \tag{2.66}$$

$$= \sum_{\text{bonds } i} \frac{k_i}{2}(b_i - b_{i,0})^2 \tag{2.67}$$

$$+ \sum_{\text{angles } i} \frac{f_i}{2}(\theta_i - \theta_{i,0})^2 \tag{2.68}$$

$$+ \sum_{\text{dihedrals } i} \frac{V_i^d}{2}[1 + \cos(n\phi)] \tag{2.69}$$

$$+ \sum_{\text{impropers } i} \kappa_i(\zeta_i - \zeta_{i,0})^2 \tag{2.70}$$

$$+ \sum_{\text{pairs } i,j} \frac{q_i q_j}{4\pi\epsilon_0\epsilon_r r_{ij}} \tag{2.71}$$

$$+ \sum_{\text{pairs } i,j} 4\epsilon_{ij}\left[\left(\frac{\sigma_{ij}}{r_{ij}}\right)^{12} - \left(\frac{\sigma_{ij}}{r_{ij}}\right)^{6}\right], \tag{2.72}$$

where k_i, f_i and $\kappa_i/2$ denote the spring constants of the harmonic potentials for the bond length b_i, angle θ_i and improper dihedral angle ζ_i, around their respective rest lengths, indicated by the subscript 0. Dihedral potentials depend on the barrier height V_i^d between different conformers and the periodicity n. Coulomb interactions at a distance r_{ij} between two atoms are calculated based on the charges q_i, the vacuum permittivity ϵ_0 and permittivity ϵ_r, whereas σ_{ij} and ϵ_{ij} denote the Lennard-Jones parameters. Note that the partial charges aim at reproducing the electrostatic potential not only for localized charges such as ions, but also for, e.g., delocalized electron systems, and, therefore, are generally non-integer values.

Developing force fields that give accurate results for a large variety of systems is a challenging task. Therefore, a number of different prominent force fields have been developed, such as the amber [206, 207], charmm [208, 209], opls [210], and gromos [211] force fields. These are often more accurate in predicting a number of experimental observables correctly, such as conformational energies [212, 213], and less accurate in predicting others. Similarly, many force fields are reasonable accurate for, e.g., globular proteins with a conserved structure, but give largely differing results for, e.g., intrinsically disordered proteins [214]. As such, to check the robustness of a prediction, it has become common practice to repeat the same simulations using different force fields [215] as well as to constantly validate these predictions through comparison with experiments.

Generally, force fields and MD simulations do not allow chemical reactions or bond breaking. Similarly, polarization effects are usually not accounted for. However, approaches into both directions have been developed [216–220] and may become more widely used in the future.

The force field mainly used in chapters 3 and 5 of this thesis is the Generalized Amber Force Field (GAFF) [221]. The reason is simply that simulation setups, which were already equilibrated with GAFF, are conveniently available from the SolvationToolkit package [222]. In these chapters, to separate force field errors from sampling errors, the accuracies of different approaches are always obtained by comparison of a large number of short simulations with respect to a converged reference result, based on the same force field. As such, the particular choice of a force field is, in this context, unimportant. Naturally, for the success of the field of computational free energy calculations as a whole, force fields are a crucial factor.

Temperature and Pressure Coupling

A canonical system is coupled to the temperature, and in most cases, also the pressure of a much larger external system. Thermostats and barostats are algorithms that model these coupling effects.

To start with one of the simplest thermostats, it is considered that the instantaneous temperature T of the system and the velocities $\{\mathbf{v}_i\}$ of n_p particles are related via

$$\frac{3}{2}k_B T = \left\langle \frac{m_i \mathbf{v}_i^2}{2} \right\rangle \tag{2.73}$$

$$= \frac{1}{n_p} \sum_{i=1}^{n_p} \frac{m_i \mathbf{v}_i^2}{2} \ . \tag{2.74}$$

If the target (= external) temperature T' differs from the current one, then the simplest approach is to rescale all velocities according to

$$\mathbf{v}_i' = \sqrt{\frac{T'}{T}}\mathbf{v}_i \tag{2.75}$$

such that T' is reached. This method is referred to as velocity rescaling. However, as the temperature is not able to fluctuate in this approach, the properties of a canonical system, which free energy energy simulations are conducted in, are not reproduced. One option to circumvent this problem is to randomly draw the components of all velocities $\mathbf{v}_i''$ from a Gaussian distribution around the ones of $\mathbf{v}_i'$, i.e.,

$$p(v_{i,x}'') = \frac{1}{\sqrt{2\pi}\sigma_i} \exp\left(-\frac{v_{i,x}'^2}{2\sigma_i^2}\right) \; , \tag{2.76}$$

with variance $\sigma_i^2 = k_B T / m_i$, where, as an example, $v_{i,x}''$ denotes the final velocity of particle i in x-direction. Conducting this procedure only after certain time intervals gives the system further room for fluctuations.

The main advantage of velocity scaling is its simplicity with respect to implementation. Whereas the resulting ensemble does not correspond exactly to the canonical one, the deviations are in practice often rather small [223]. It was therefore used, e.g., for the simple test system of a LJ gas with a self-written simulation program in chapter 3. A more refined method yielding more continuous velocities as well as better ensemble properties is the Nose-Hoover thermostat [224, 225]. Here, an additional degree of freedom γ is introduced that determines an (anti)friction term in the equations of motion

$$\frac{\mathrm{d}}{\mathrm{d}t} m_i \mathbf{v}_i = \mathbf{F}_i - 2\gamma \mathbf{v}_i \tag{2.77}$$

$$\frac{\mathrm{d}\gamma}{\mathrm{d}t} = \frac{1}{\tau}\left(\sum_i \frac{1}{2} m_i \mathbf{v}_i^2 - \frac{3}{2} n_p k_B T\right) \; , \tag{2.78}$$

where τ controls the timescale of the temperature regulation, and $\mathbf{F}_i$ denotes the total force acting on particle i. The Nose-Hoover thermostat is used in all simulations based on the GROMACS software package in this thesis.

Similar principles apply to barostats that regulate the pressure for NPT ensembles. Similar to velocity rescaling, the Berendsen barostat [223] rescales the positions, and thereby the volume of the system. The rescaling is, however, not conducted in one step; instead it depends on a parameter τ_p that regulates how fast the target pressure should be reached. The advantage of the Berendsen barostat is that it is well behaved, and therefore often used to stabilize a system prior to further equilibration and production runs. However, similarly to velocity-rescaling it does not produce the correct (isobaric-isoenthalpic) canonical ensembles.

The Anderson barostat [226] uses the volume as an additional degree of freedom, which is coupled to the system through equations of motion, similar to a piston. The volume is therefore assigned a kinetic and potential energy, where the latter is determined through the difference of the instantaneous to the target (= external) pressure. A conceptual "mass" is assigned to the volume, which is a user-chosen parameter that regulates the extent of the system fluctuations. The Parrinello-Rahman [227, 228] barostat is an extension that allows the shape of the volume to deviate from the one it was initially assigned to. Both barostats correctly reproduce the canonical ensemble.

In this thesis, if not specified otherwise, all GROMACS based simulations follow the same procedure: First, an energy minimization is conducted of the system, using the steepest–descent algorithm [229]. Secondly, the system is equilibrated in an NVT ensemble, followed by equilibration in an NPT ensemble. The latter is divided in two parts: Firstly, using the Berendsen barostat, secondly, using the Parrinello-Rahman barostat. All subsequent production runs (i.e., the parts of the simulations that are used to calculate the free energy differences) are conducted with the Parrinello-Rahman barostat.

3

Variationally Derived Intermediates

3.1 Determining Free-Energy Differences Through Variationally Derived Intermediates

The following section consists of the article

M. Reinhardt, H. Grubmüller, Determining Free-Energy Differences Through Variationally Derived Intermediates, Journal of Chemical Theory and Computation vol. 16, issue 6, pp. 3504-3512 (2020). [230]

The format, including the numbering of equations, figures and tables, has been altered to match the format of this thesis. Furthermore, all references are listed in the bibliography at the end of this thesis rather than at the end of this article.

Both authors contributed to conceiving the study and writing the manuscript. I implemented, conducted and analyzed all test systems and simulations.

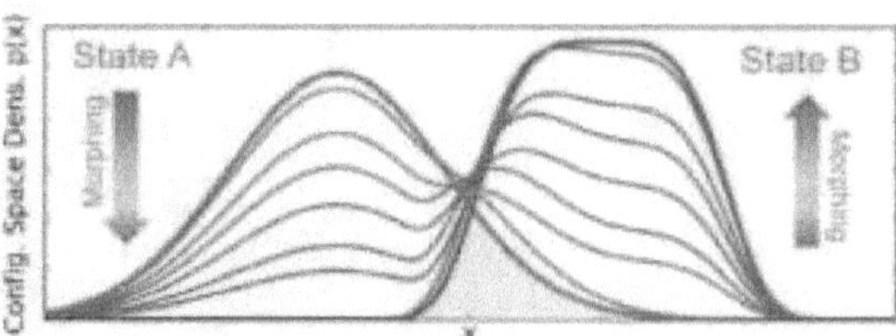

Figure 3.1: Table of Contents Image

3.2 Introduction

Free energy calculations provide essential insights into numerous physical and biochemical systems. Examples range from predicting binding processes of biomolecules for drug design [8, 10, 11] to determining thermodynamic properties of crystalline materials [13, 14]. For large and complex systems with slow relaxation rates and typically 10^5 to 10^7 particles, only limited accuracy is achieved [231], despite sub-stantial methodological progress [84, 183, 232, 233] and immense computational effort. Besides force field inaccuracies, insufficient sampling is the main bottle-neck [74]. Within a generalized framework connecting two of the most established methods, Free Energy Perturbation (FEP) [79] and the Bennett Acceptance Ra-tio method (BAR) [80], with the latter generally considered the more accurate one [171], we here will develop and evaluate a variational approach for optimal sampling that minimizes the error due to limited sampling.

Given the Hamiltonians $H_1(\mathbf{x})$ and $H_N(\mathbf{x})$ of two states 1 and N, where $\mathbf{x} \in \mathbb{R}^{3M}$ denotes the position of all M particles of the simulation system, the free energy difference $\Delta G_{1,N}$ between these states is given by the Zwanzig formula [79],

$$\Delta G_{1,N} = -\ln\langle e^{-[H_N(\mathbf{x})-H_1(\mathbf{x})]}\rangle_1 . \tag{3.1}$$

where $\langle\rangle_1$ denotes an ensemble average defined by $H_1(\mathbf{x})$, which is approximated by averaging over a finite sample of size n obtained from atomistic simulations or Monte Carlo sampling. For ease of notation, all energies are expressed in units of $k_B T$.

Alchemical transformations substantially reduce errors in the free energy estimates [184, 185] by introducing $N-2$ intermediate states s and accumulating small free energy differences between all adjacent states s and $s+1$,

$$\Delta G_{1,N} = \sum_{s=1}^{N-1} \Delta G_{s,s+1} . \tag{3.2}$$

Using the Zwanzig formula between s and $s+1$, this technique is referred to as FEP. The same approach is also employed in other fields, for example in the context of Bayesian statistics, where the plausibility of two different models is compared by calculating their marginal likelihood ratio [78, 139].

The most common interpolation scheme for the intermediate states is along a path variable λ

$$H_s(\mathbf{x}) = (1-\lambda_s)H_1(\mathbf{x},\lambda_s) + \lambda_s H_N(\mathbf{x},\lambda_s), \quad \lambda_s \in [0,1] . \tag{3.3}$$

Figure 3.2(a) shows as a simple example a linear interpolation between two one-dimensional Hamiltonians $H_1(\mathbf{x})$ and $H_{N=9}(\mathbf{x})$. In the case of soft-core potentials [91–93], a non-linear dependence of the end states $H_1(\mathbf{x},\lambda)$ and $H_N(\mathbf{x},\lambda)$ on λ is introduced under the requirement that $H_1(\mathbf{x},\lambda=0)=H_1(\mathbf{x})$ and $H_N(\mathbf{x},\lambda=1)=H_N(\mathbf{x})$. In this context, it has been attempted to find better sequences of Hamiltonians by optimizing the distribution of λ points for a given form of a sequence or pathway[165].

Even though there is some freedom in the construction of these transformation sequences, Eq. (3.3) describes only a very small subset of all possible sequences of intermediate states, and in this sense, is not the most general. Specifically, the terms containing the information and parameters of the two different end states are always combined in an additive manner, and, e.g., any definition of intermediate Hamiltonians $H_s(\mathbf{x})$ that would involve cross terms of the form

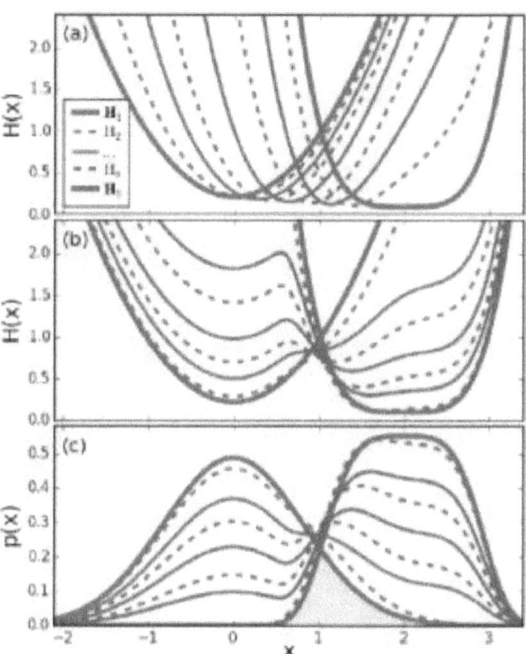

Figure 3.2: Sequences of intermediates between a harmonic potential $H_1(\mathbf{x}) = \frac{1}{2}x^2 + b$ and a quartic potential $H_9(\mathbf{x}) = (x - x_0)^4 + c$ (thick lines), where b and c have been determined such that $Z_1 = Z_9 = 1$, i.e., $\Delta G_{1,9} = 0$. Sampling states are described by odd-numbered Hamiltonians (solid lines), virtual target states by even-numbered ones (dashed lines). (a) A linear interpolation between $H_1(\mathbf{x})$ and $H_9(\mathbf{x})$. For better visualization, the intermediates are vertically offset to align the minima. (b) Intermediate Hamiltonians and (c) resulting configuration space densities of VI. The yellow area highlights the configuration space density overlap K between states 1 and 9.

$f(H_1(\mathbf{x}, \lambda) \cdot H_N(\mathbf{x}, \lambda))$ would not be possible with the definition of Eq. (3.3).

Therefore, a number of alternative approaches, modifying Eq. (3.3), have been developed. For example, an empirical potential has been proposed in the Enveloping Distribution Sampling (EDS) method [100, 101] that interpolates between the configuration space densities (rather than the Hamiltonians) of two or several end states and is, as we will find, remarkably close to the optimal solution for a single intermediate state.

Further, a continuous path between two such end states, the 'minimum variance path' (MVP), which optimizes the variance for Thermodynamic Integration [90] (TI) was derived by Blondel [194] and later, through a different formalism, by Pham and Shirts [105] based on the results from the statistical sciences by Gelman and Meng [78].

Here, we will generalize this interpolation scheme for FEP and BAR. Specifically, we ask which sequence $H_2(\mathbf{x}) \dots H_{N-1}(\mathbf{x})$ — amongst *all possible* sequences of higher order functions $\{H_s|H_1(\mathbf{x}), H_N(\mathbf{x})|\}$ that map two functions onto functions H_s (with $s = 2, \dots, N-1$) — yields, on average, the smallest mean squared error (MSE),

$$\mathrm{MSE}\left(\Delta G^{(n)}\right) = \mathbb{E}\left[\left(\Delta G - \Delta G^{(n)}\right)^2\right], \tag{3.4}$$

of the free energy estimate $\Delta G^{(n)}$ obtained through finite sampling with n sample points with respect to the *exact* free energy difference ΔG. Figures 3.2(b) and (c) show such a general interpolation sequence, which we refer to as *Variationally-derived Intermediates* (VI) method.

Our approach differs from previous approaches in that here we optimize the full MSE. Because the MSE can be decomposed into the sum of variance, $\mathbb{E}\left[\left(\mathbb{E}\left[\Delta G^{(n)}\right] - \Delta G^{(n)}\right)^2\right]$ and squared bias, $\left(\mathbb{E}\left[\Delta G - \Delta G^{(n)}\right]\right)^2$, it has been attempted to analyze and optimize these two terms separately [105, 171, 234, 235]. For the MVP in the context of TI, continuous sampling along the path variable λ is assumed; for practical applications, however, discrete integration is preferred, which implies an additional bias that is difficult to assess and, therefore, not included within the optimization. As we will find, optimizing the sum of both, variance and bias, yields a conceptionally improved result.

3.3 Theory

Optimal Mean Squared Error Sequence of Intermediates for Free Energy Perturbation

To solve the above variational problem and to find the optimal sequence of H_s, we consider the FEP scheme displayed in Fig. 3.3(a) as one of several possible implementation of Eq. (3.2) using Eq. (3.1). In this particular variant, which is symmetric with respect to exchange of the two end states to avoid hysteresis effects, sample points are solely drawn from the odd-numbered 'sampling states', indicated by the solid lines in Figs. 3.2 and 3.3. The even-numbered states serve as virtual 'target states' (dashed lines), similar to e.g. the Overlap Sampling method [236]. From the sum of the individual perturbation steps, the average MSE of this scheme is

$$\mathrm{MSE}\left(\Delta G_{1,N}^{(n)}\right) = \mathbb{E}\left[\left(\Delta G_{1,N} - \sum_{\substack{s=1 \\ s\,\mathrm{odd}}}^{N-2}\left(\Delta G_{s\to s+1}^{(n)} - \Delta G_{s+2\to s+1}^{(n)}\right)\right)^2\right]. \tag{3.5}$$

As in Fig. 3.3, the arrows point from sampling to target states.

Assuming for each sample state s a set of n independent sample points $\{\mathbf{x}_i\}$, drawn from $p_s(\mathbf{x}) = e^{-H_s(\mathbf{x})}/Z_s$, with partition function Z_s, the terms arising from expanding Eq. (3.5),

$$\mathrm{MSE}\left(\Delta G_{1,N}^{(n)}\right) = (\Delta G_{1,N})^2 + \sum_{\substack{s=1 \\ s\,\mathrm{odd}}}^{N-2} \mathbb{E}\left[\left(\Delta G_{s\to s+1}^{(n)}\right)^2 + \left(\Delta G_{s+2\to s+1}^{(n)}\right)^2\right]$$

$$- 2\Delta G_{1,N}\left(\sum_{\substack{s=1 \\ s\,\mathrm{odd}}}^{N-2}\left(\mathbb{E}\left[\Delta G_{s\to s+1}^{(n)}\right] - \mathbb{E}\left[\Delta G_{s+2\to s+1}^{(n)}\right]\right)\right) \tag{3.6}$$

$$- \sum_{\substack{s=1 \\ s\,\mathrm{odd}}}^{N-2}\sum_{\substack{t=1 \\ t\,\mathrm{odd}}}^{N-3} \mathbb{E}\left[2\,\Delta G_{s\to s+1}^{(n)}\,\Delta G_{t+2\to t+1}^{(n)}\right]$$

will be considered one by one. As the exact free energy difference is a constant,

$$\mathbb{E}\left[\Delta G_{1,N}\right] = \Delta G_{1,N}. \tag{3.7}$$

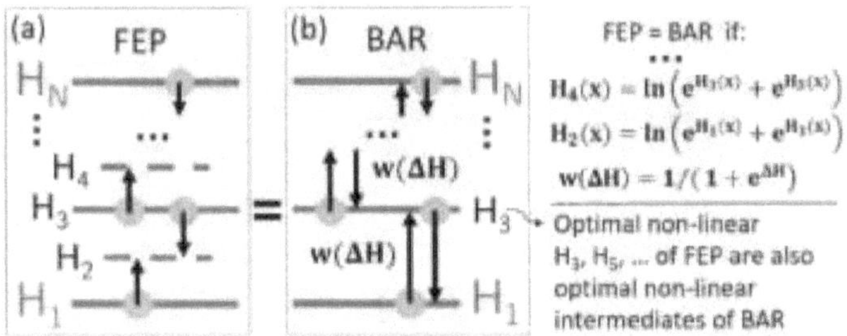

Figure 3.3: Two schemes of free energy calculation. Yellow dots represent sample sets in the respective potential; arrows indicate the evaluation of differences $\Delta H(x)$ between adjacent Hamiltonians. Free energy differences are either determined by (a) FEP, or by (b) BAR with multiple steps. Both schemes give identical results at the stated conditions.

For the linear term, the average over all sample realizations reads

$$\mathbb{E}\left[\Delta G^{(n)}_{s\to s+1}\right] = -\int p_s(\mathbf{x}_1)d\mathbf{x}_1 \ldots \int p_s(\mathbf{x}_n)d\mathbf{x}_n$$
$$\ln\left[\frac{1}{n}\sum_{i=1}^{n} e^{-(H_{s+1}(\mathbf{x}_i)-H_s(\mathbf{x}_i))}\right],$$
(3.8)

and for the quadratic term

$$\mathbb{E}\left[\left(\Delta G^{(n)}_{s\to s+1}\right)^2\right] = \int p_s(\mathbf{x}_1)d\mathbf{x}_1 \ldots \int p_s(\mathbf{x}_n)d\mathbf{x}_n$$
$$\left(\ln\left[\frac{1}{n}\sum_{i=1}^{n} e^{-(H_{s+1}(\mathbf{x}_i)-H_s(\mathbf{x}_i))}\right]\right)^2.$$
(3.9)

Similar expressions are obtained for $\Delta G^{(n)}_{s+2\to s+1}$. The exact free energy differences are

$$\Delta G_{s,s+1} = -\ln\int e^{-(H_{s+1}(\mathbf{x})-H_s(\mathbf{x}))}p_s(\mathbf{x})d\mathbf{x}.$$
(3.10)

For shifted Hamiltonians $H'_s(\mathbf{x}) = H_s(\mathbf{x}) - C_s$ and $H'_{s+1}(\mathbf{x}) = H_{s+1}(\mathbf{x}) - C_{s+1}$, Eq. (3.1) yields

$$\Delta G^{(n)}_{s'\to(s+1)'} = \Delta G^{(n)}_{s\to s+1} - C_{s+1} + C_s,$$
(3.11)

which also holds for the exact value $\Delta G_{1',N'}$. The offsets on the right hand side of Eq. (3.11) cancel out when calculating the MSE of Eq. (3.5). Choosing C_s and

C_{s+1} such that the term in the logarithm of Eqs. (3.8) and (3.9) is close to one, and thus all $\Delta G^{(n)}_{s' \to (s+1)'}$ are small with respect to $k_B T = 1$, first order expansion of the logarithm allows to factorize the integrals, therefore,

$$\mathbb{E}\left[\Delta G^{(n)}_{s' \to (s+1)'}\right] = -\int e^{-(H'_{s+1}(\mathbf{x}) - H'_s(\mathbf{x}))} p_s(\mathbf{x})\, d\mathbf{x} + 1 . \tag{3.12}$$

For the shifted Hamiltonians, the same expansion can be applied to the exact free energy difference of Eq. (3.10). Therefore, Eq. (3.12) reduces to

$$\mathbb{E}\left[\Delta G^{(n)}_{s' \to (s+1)'}\right] = \Delta G_{s', s'+1} . \tag{3.13}$$

The configuration space densities of the shifted and the initial Hamiltonians are identical, i.e.

$$p'_s(\mathbf{x}) = \frac{e^{-H_s(\mathbf{x}) - C_s}}{\int e^{-H_s(\mathbf{x}) - C_s}\, d\mathbf{x}} = p(\mathbf{x}). \tag{3.14}$$

Note the underlying assumption that the same offsets C_s and C_{s+1} can be used to enable the series expansion of the exact and the estimated free energy difference. This assumption, as we will later find, holds for the vast majority of cases, but may break down in case of very few sample points and very low configuration space density overlap between two neighboring states.

For the cross terms in Eq. (3.6), note that the estimated free energy differences of the individual steps are based on uncorrelated sample sets, and therefore

$$\mathbb{E}\left[\Delta G^{(n)}_{s' \to t'} \cdot \Delta G^{(n)}_{u' \to v'}\right] = \mathbb{E}\left[\Delta G^{(n)}_{s' \to t'}\right] \mathbb{E}\left[\Delta G^{(n)}_{u' \to v'}\right]$$
$$= \Delta G_{s', t'}\, \Delta G_{u', v'} , \tag{3.15}$$

for $(s' \to t') \neq (u' \to v')$. Expanding Eq. (3.9) yields

$$\mathbb{E}\left[\left(\Delta G^{(n)}_{s' \to (s+1)'}\right)^2\right] = \frac{1}{n^2} \sum_{i=1}^{n} \int p_s(\mathbf{x}_1)\, d\mathbf{x}_1 \cdots \int p_s(\mathbf{x}_n)\, d\mathbf{x}_n$$
$$\left(e^{-(H'_{s+1}(\mathbf{x}_i) - H'_s(\mathbf{x}_i))} - 1\right)^2$$
$$+ \frac{1}{n^2} \sum_{i=1}^{n} \sum_{\substack{j=1 \\ j \neq i}}^{n} \int p_s(\mathbf{x}_1)\, d\mathbf{x}_1 \cdots \int p_s(\mathbf{x}_n)\, d\mathbf{x}_n$$
$$\left(e^{-(H'_{s+1}(\mathbf{x}_i) - H'_s(\mathbf{x}_i))} - 1\right)\left(e^{-(H'_{s+1}(\mathbf{x}_j) - H'_s(\mathbf{x}_j))} - 1\right) . \tag{3.16}$$

Using Eq. (3.15), all expressions from the cross terms only depend on exact free energy differences. Summarizing these terms by $f_{s'}$, Eq. (3.16) can be written as

$$\mathrm{E}\left[\left(\Delta G_{s'\to(s+1)'}^{(n)}\right)^2\right] = \frac{1}{n}\int e^{-2(H_{s+1}'(\mathbf{x})-H_s'(\mathbf{x}))}p_s(\mathbf{x})\mathrm{d}\mathbf{x} \\ + f_{s'}(\Delta G_{s',(s+1)'}).$$

(3.17)

Inserting Eqs. (3.13) and (3.17) into Eq. (3.5),

$$\mathrm{MSE}\left(\Delta G_{1,N}^{(n)}\right) = \sum_{\substack{s=1 \\ s\,\mathrm{odd}}}^{N-2} \frac{1}{n}\Bigg(\int p_s(\mathbf{x})\,\mathrm{d}\mathbf{x}\, e^{-2(H_{s+1}'(\mathbf{x})-H_s'(\mathbf{x}))} \\ + \int p_{s+2}(\mathbf{x})\,\mathrm{d}\mathbf{x}\, e^{-2(H_{s+1}'(\mathbf{x})-H_{s+2}'(\mathbf{x}))} \\ + g_{s'}\left(\Delta G_{s',(s+1)'},\Delta G_{(s+2)',(s+1)'},\Delta G_{1',N'}\right)\Bigg),$$

(3.18)

where $g_{s'}$ again denotes an expression that only depends on exact free energy differences and thus is dropped for the optimization below.

3.4   Results and Discussion

Optimal Sequence

With these expressions, the variational problem can be solved analytically. For the odd-numbered states s, variation of $\mathrm{MSE}\left(\Delta G_{1,N}^{(n)}\right)$, Eq. (3.18),

$$\frac{\partial}{\partial H_s(\mathbf{x})}\left(\mathrm{MSE}\left(\Delta G_{1,N}^{(n)}\right) + \nu\int(e^{-H_s(\mathbf{x})} - Z_s)\mathrm{d}\mathbf{x}\right) \overset{!}{=} 0$$

(3.19)

yields

$$H_s(\mathbf{x}) = -\frac{1}{2}\ln\left(e^{-2(H_{s-1}(\mathbf{x})-C_{s-1})} + e^{-2(H_{s+1}(\mathbf{x})-C_{s+1})}\right),$$

(3.20)

where $Z_s = \int e^{-H_s(\mathbf{x})}\mathrm{d}\mathbf{x}$ is the (finite) partition sum and ν is a Lagrange multiplier.

Similarly, for the even-numbered states,

$$H_s(\mathbf{x}) = \ln\left(e^{H_{s-1}(\mathbf{x})-C_{s-1}} + e^{H_{s+1}(\mathbf{x})-C_{s+1}}\right).$$

(3.21)

An additive term C_s in Eqs. (3.20) and (3.21) was omitted, as it cancels in $\Delta G_{s-1\to s}^{(n)} - \Delta G_{s+1\to s}^{(n)}$. The result is a set of equations for all states s for which each Hamiltonian $H_s(\mathbf{x})$ depends only on the two adjacent states. The initial requirement for small $\Delta G^{(n)}_{s' \to (s+1)'}$ is fulfilled by setting $C_s = -\ln Z_s$, as in this case, all Z'_s are one. Rearranging terms for *odd* s,

$$e^{-2H_s(\mathbf{x})} = e^{-2H_{s-1}(\mathbf{x})} \cdot r_{s-1,s}^{-2} + e^{-2H_{s+1}(\mathbf{x})} \cdot r_{s+1,s}^{-2} \tag{3.22}$$

and for the virtual target states, i.e. even s,

$$e^{H_s(\mathbf{x})} = e^{H_{s-1}(\mathbf{x})} \cdot r_{s-1,s} + e^{H_{s+1}(\mathbf{x})} \cdot r_{s+1,s} \tag{3.23}$$

with $r_{s,t} = Z_s/Z_t$.

Eqs. (3.22) and (3.23) are the first main result of this article, they define the sequence of Hamiltonians yielding the best MSE for FEP free energy calculations.

Generalization of BAR

The second main result is that Eq. (3.21) serves to generalize the BAR formula. To see this, consider Eq. (3.21) for $N = 3$, i.e., with one intermediate target state. Applied to the two involved free energy differences, the Zwanzig formula yields

$$\Delta G^{(n)}_{1,3} = \Delta G^{(n)}_{1 \to 2} - \Delta G^{(n)}_{3 \to 2} \tag{3.24}$$

$$= -\ln\langle e^{-[H_2(\mathbf{x}) - H_1(\mathbf{x})]}\rangle_1 + \ln\langle e^{-[H_2(\mathbf{x}) - H_3(\mathbf{x})]}\rangle_3. \tag{3.25}$$

Inserting Eq. (3.21) as the target state Hamiltonian $H_2(\mathbf{x})$ yields the BAR formula

$$e^{-(\Delta G_{1,2} - C)} = \left\langle \frac{1}{1 + e^{H_3(\mathbf{x}) - H_1(\mathbf{x}) - C}} \right\rangle_1 \Big/ \left\langle \frac{1}{1 + e^{H_1(\mathbf{x}) - H_3(\mathbf{x}) + C}} \right\rangle_3, \tag{3.26}$$

with $C = C_3 - C_1$.

Notably, the above derivation yields the more general result that Eq. (3.26) provides the best MSE free energy estimate also for finite and small n, even down to $n = 1$ given sufficient configuration space density overlap between adjacent states, which is fulfilled, for instance, in the limit of many intermediates. In contrast, because the derivation by Bennett [80] strictly holds only for infinite sampling, so far n was required to be large, and proper convergence had to be assumed. Further, in the original derivation [80] the error distribution of the free energy estimates had to be assumed to be Gaussian, which in our above result is also not required. While it has been known that BAR yields the lowest variance out of the asymptotically unbiased estimators[166], the above derivation shows that this also holds for the MSE at finite n, and that BAR is the best out of all estimators, including, in addition, also the asymptotically biased ones. In the context of the Overlap Sampling method [184, 185, 187, 189, 236], it has been shown that a virtual FEP intermediate can be defined that yields the weighting function from Bennett's derivation; the above results prove that this intermediate is indeed optimal for the FEP scheme. Note that, in the extreme case of small configuration space density overlap and very few sample points, the average deviation between the series expansions of the exact and the estimated expressions of Eq. (3.12) can become too large, in which case our approach may miss the absolute optimum and, therefore, a better solution may exist. However, as we will see later in the context of Fig. 3.7, the VI result yields a better MSE than all other approaches that we assessed, even for small n at small configuration space density overlaps between the end states.

The third main result is that the optimal intermediates for FEP are also optimal for BAR. To see this, consider again the Eqs. (3.22) and (3.23) yielding optimal FEP intermediates for any (odd) number $N - 2$ of intermediate states. As was shown in the derivation of Eq. (3.26), using the intermediate of Eq. (3.23) with FEP between two sampling states is equivalent to using BAR between these two. Applied recursively to many states, and as illustrated in Fig. 3.3, the $\tilde{N} = (N+1)/2$ sampling states from any sequence of N FEP-optimal Hamiltonians $\{H_s(\mathbf{x})\}$ are also optimal for BAR with multiple states, where so far, too, mostly states have been used of the form as in Eq. (3.3). The governing system of equations for BAR with multiple states is obtained by replacing $H_{s-1}(\mathbf{x})$ and $H_{s+1}(\mathbf{x})$ in Eq. (3.22) with the expression of Eq. (3.23), yielding for odd s

$$e^{-2H_s(\mathbf{x})} = \left(e^{H_{s-2}(\mathbf{x})} r_{s-2,s} + e^{H_s(\mathbf{x})} \right)^{-2} + \left(e^{H_{s+2}(\mathbf{x})} r_{s+2,s} + e^{H_s(\mathbf{x})} \right)^{-2} . \tag{3.27}$$

Here, the sampling states are now coupled directly, and only knowledge of the partition sum ratios between these is required. Solving the system of equations of Eq. (3.27) for all sampling states yields the intermediates with optimal MSE for BAR. Conversely, for the setup of one sampling state between two given target end states 1 and 3, Eq. (3.20) recovers the EDS potential when using a factor of 2 in the exponent of Ref. [101]. In summary, both BAR and EDS are special cases of, or approximations to, our more general variational VI result that also requires fewer assumptions.

To solve Eqs. (3.22) and (3.23) for the optimal FEP intermediate Hamiltonians $H_s(\mathbf{x})$, or, alternatively, Eq. (3.27) to directly obtain the optimal BAR intermediates, respectively, note that the unknown free energy differences $\Delta G_{s,t} = -\ln r_{s,t}$ are part of the equations which, therefore, have to be solved iteratively. With an initial guess for all $r_{s,t}$, the set of equations is solved in a point-wise fashion for any given $\mathbf{x}$. After sampling all odd-numbered states, the $r_{s,t}$ values are updated iteratively, such that the sequence of intermediate states converges towards the optimum. This iteration converges to the optimal result, because both, the estimates as well as the linear approximation of the series expansion in Eq. (3.12) converge simultaneously. For a typical biomolecular many-body system, the additional computational effort is small compared to computing $H_1(\mathbf{x})$ and $H_N(\mathbf{x})$.

For the above illustrative example, Fig. 3.2(b) and (c) show the optimized Hamiltonians and the configuration space densities, respectively, of the converged sequence of intermediate states. To this end, initial values $r_{s,t} = 1$ were used and Eqs. (3.22) and (3.23) were iterated until convergence, using numerical integration over $\mathbf{x}$ and updating the $r_{s,t}$ during the process. Unlike the linear interpolations shown in Fig. 3.2(a), the VI sequence leads to a probability density, which gradually decreases in the region of A and increases in the region of B, while remaining almost constant at the point of maximum configuration space overlap.

One-dimensional Test Case

Figure 3.4(a) shows the results of numerical simulations using the one-dimensional test case shown in Fig. 3.2. Different equilibrium constants x_0 (42 different values) are used for $H_N(\mathbf{x})$, thereby varying configuration space overlaps

$$K = \int_{-\infty}^{\infty} \min(p_1(\mathbf{x}), p_N(\mathbf{x})) d\mathbf{x} \tag{3.28}$$

between the end states, indicated by the yellow area in Fig. 3.2(b). Sets of $n = 100$ uncorrelated sample points are drawn from $p_s(\mathbf{x})$ through rejection sampling in each of the $\tilde{N} = 3$ sampling states. Based on these sets, BAR (solid lines) is used to calculate the free energy difference between the individual states, and subsequently, between the start and end state. As a comparison, the dashed lines in Fig. 3.2(a) show the results using MBAR[166], where the differences in the Hamiltonians for all states are considered for each sample point. The free energy estimate is compared to the exact free energy difference. For each K, the MSE,

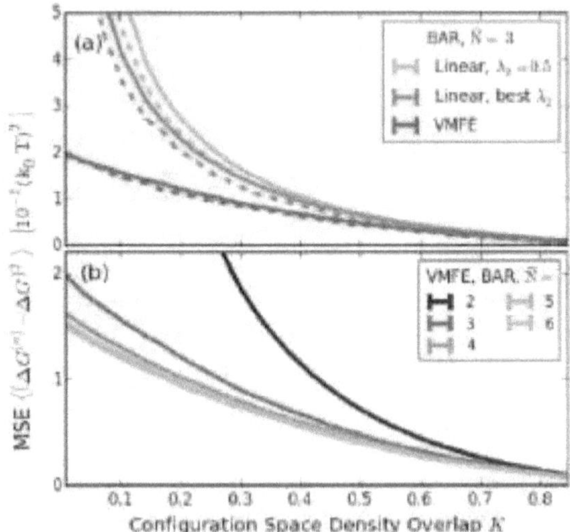

Figure 3.4: Accuracies of free energy calculations for different overlaps between the end states, determined numerically for the model Hamiltonians from Fig. 3.2. For the solid lines, BAR is used between adjacent sampling states, for the dashed lines, MBAR is used. (a) Comparison between VI and two variants of linear intermediates: a linearly spaced λ_2 and an empirically optimized λ_2 yielding the lowest MSE. (b) Accuracies for different numbers of VI sampling states for a given total sampling size.

Eq. (3.5), is calculated by averaging over 600,000 of such realizations.

VI (blue curve) yields the smallest MSD for all K, compared to both the first linear interpolation variant (light green) using a linearly spaced $\lambda_2 = \frac{1}{2}$, like in a typical free energy calculation, and even compared to the second variant (dark green) using the empirically determined λ_2 value that yields the best MSE that can be achieved by linear interpolation. To obtain the latter, we loop over the allowed range between zero and one in steps of 0.01. To reliably calculate the MSD with respect to the exact value, for each λ_2, 150,000 free energy estimates are calculated. Once the lowest MSE λ_2 is determined, the corresponding MSD is calculated once again using 600,000 realizations. The procedure is repeated for each value of K. We note that the λ_2 yielding the best MSE varies for different K, and is inaccessible in practice for high-dimensional systems.

The largest improvements of VI are seen for small configuration space density overlaps that notoriously cause the largest uncertainties. Also for MBAR[166], shown by the dashed lines in Fig. 3.4(a), VI gives the better MSE than the linear intermediates. Figure 3.4(b) shows how the MSE of VI improves with increasing number of states $\tilde{N}$, keeping the total number of sample points, and hence the total computational effort, constant. For this example, the MSE increases up to $\tilde{N} = 5$, beyond which no further improvement appears.

Approximated Sequence and Comparisons

In the above VI scheme, Eqs. (3.22) and (3.23) connect all intermediates and, therefore, cannot be solved efficiently in a straightforward way for many-particle systems. To overcome this limitation, we propose and assess two approximations which will yield analytical expressions for H_a that can be used even in large scale simulations. The first approximation is to switch to a hierarchical solution for the VI scheme. In a first step, the sampling state in the middle of the sequence, $\hat{H}_{\tilde{N}/2}$, is determined as the optimal state between H_1 and $H_{\tilde{N}}$ using Eq. (3.22). In the next step, $\hat{H}_{\tilde{N}/4}$ is determined as the optimal state between H_1 and $\hat{H}_{\tilde{N}/2}$ as well as $\hat{H}_{3\tilde{N}/4}$ between $\hat{H}_{\tilde{N}/2}$ and $H_{\tilde{N}}$, and so on. The hat above the Hamiltonian indicates the approximated form.

The second approximation is that only the sampling states are coupled using Eq. (3.22), and no virtual states are used. Therefore, while still using BAR between two adjacent sampling states, the states are optimized as if the Zwanzig formula, i.e., Eq. (3.1) was used to calculate the free energy difference between them.

Using these two approximations, an analytical result for the sequence of intermediate states is obtained.

$$\hat{H}_a(\mathbf{x}) = -\frac{1}{2} \ln\left[(1 - \zeta_a)e^{-2H_1(\mathbf{x})} + \zeta_a e^{-2(H_{\tilde{N}}(\mathbf{x}) - C)} \right], \tag{3.29}$$

where all $\hat{H}_a(\mathbf{x})$ are a function of only H_1 and $H_{\tilde{N}}$ and $\zeta \in [0, 1]$, and only $C = C_N - C_1 \approx \Delta G$ has to be determined iteratively. Consequently, no other prior knowledge of the differences between the individual states is required, and therefore, the sampling simulations for each state can be run in parallel without communication. Note that these two approximations introduce a parameter ζ_a.

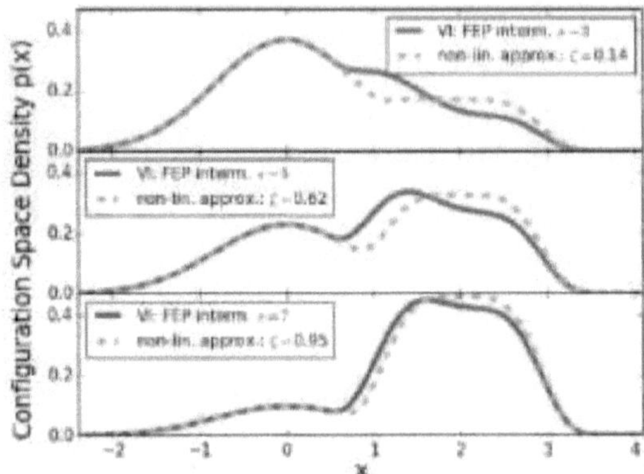

Figure 3.5: Comparison between configuration space densities of the approximated VI sequence, i.e., Eq. (3.29) (dashed lines) with that of the optimal VI sequence (solid lines) for the one-dimensional example shown in Fig. 3.2(c). For better visualization, the three intermediate sampling states $s = 3, 5, 7$ are shown separately.

which is not part of the exact result, Eqs. (3.22) and (3.23), and here plays a similar role as the λ_s in Eq. (3.3) for the linear intermediates.

For the one-dimensional example in Fig. 3.2(c), Figure 3.5 compares the configuration space densities $p(x)$ of the approximate intermediate Hamiltonians $\hat{H}_s(\mathbf{x})$ (dashed lines) with those of the optimal $H_s(\mathbf{x})$ (solid lines). As can be seen, the overall shape is similar.

The approximated VI in Eq. (3.29) is similar but not identical to the form of the MVP. The obvious difference to the approximated VI is the prefactor in the exponentials (2 and 1/2, respectively). The deeper conceptual difference is, as outlined in the introduction, that the approximated VI is an approximation to the sequence that optimizes the MSE, and thereby also accounts for the biases. It is, further, optimized for FEP and BAR, explicitly considering discrete states with finite sampling. In contrast, the MVP optimizes the variance for TI in the large sample limit assuming independent samples, continuously drawn along the path variable λ.

Next, we compare both the optimal VI and the approximated VI to the

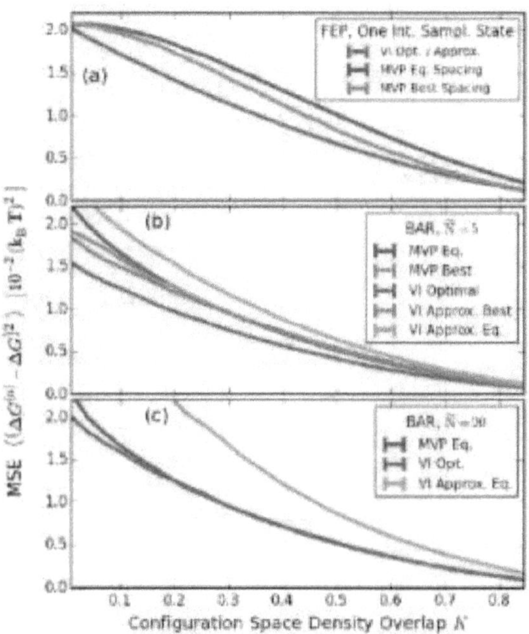

Figure 3.6: Comparison between optimal VI, approximated VI, and the MVP (Minimum Variance Path)[104, 105] at different numbers of sampling states $\tilde{N}$. (a) One intermediate sampling state, FEP is used to determine the free energy difference to the end states. In this case, optimal VI equals the mid-point of the approximated VI. (b) 5 sampling states. BAR is used to calculate the free energy difference between these. The overall number of sample points is kept constant, i.e., the number of samples per state is lower for a higher number of states. (c) Using 20 sampling states, otherwise as (b).

MVP. Figure 3.6 shows the results for different numbers of intermediates states. The same model system and procedure as the one used to obtain the results for the comparison to the linear intermediates, shown in Fig. 3.4 and described in section 3.5, is used. Again, for a fair comparison, the overall number of sample points was kept constant, i.e., the number of points per state is smaller for a larger number of intermediates.

As can be seen, for all values of K the optimal VI yields the smallest MSEs at all numbers of intermediates. The approximated VI, that is equivalent to the optimal VI at $\zeta = 0.5$ when sampling in only one intermediate state, Fig. 3.6(a), therefore also yields a smaller MSE than both the midpoint MVP ($\lambda = 0.5$, purple) and the MVP with the best lambda (light red). The latter was again determined empirically by iterating over all possible lambda values in steps of 0.01.

Conversely, for 20 states, shown in Fig. 3.6(c), the approximated VI yields a higher MSE, indicating that for increasing numbers of steps the approximations have a larger effect. Except for low values of K, similar MSEs are obtained for both the optimal VI and MVP, because for large numbers of sampling states, FEP and TI become equivalent.

Interestingly, these MSEs are larger than for optimal VI with 5 states, shown in Fig. 3.6(b), suggesting that there is an optimal number of states. Here, the MVP performs better than the approximated VI at equidistant spacing, whereas for the best spacing, the respective MSEs are similar. For five states, the best spacing was obtained by adjusting the λ and ζ values, such that the best fit of the configuration space densities with optimal VI was obtained, as shown in Fig. 3.5. Through further variation, we tested if an even better combination of path variables could be found, which was not the case.

The fact that the approximated VI and the MVP converge to the optimal solution in opposite cases indicates that they are limiting cases to the general optimal VI result.

Figure 3.7 shows how the MSE depends on the number of sample points. The same procedure as for Fig. 3.6(b) with five sampling states was used, but now with only one equilibrium distance between the minima of the harmonic and quartic potential of $x_0 = 3$ (i.e., $K \approx 0.02$, see Fig. 3.2). To avoid the problem of the optimization of a path variable, we compare only the optimal VI and the MVP

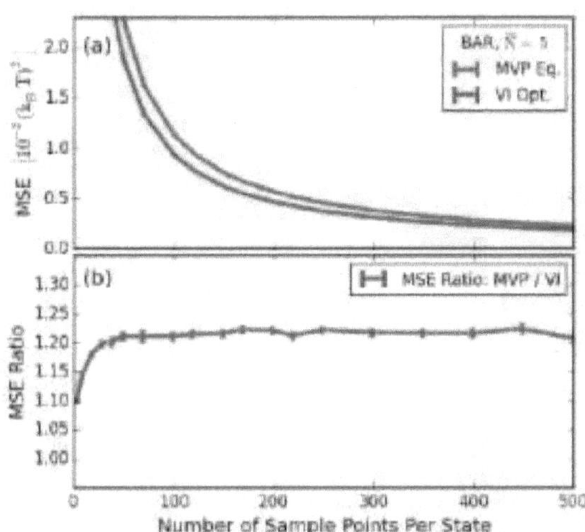

Figure 3.7: Comparison of the achieved accuracy by the minimum variance path method (MVP) and VI for different numbers of sample points per state using 5 states. (a) Obtained MSE for both methods and (b) their respective ratios.

with equidistant spacing, that, in the case of Fig. 3.6(b), yielded similar MSEs as the MVP with the best spacing.

As can be seen in panel (a), a smaller MSE is achieved by VI across a broad range of numbers of sample points. The ratio of the MSEs of both methods, Fig. 3.7(b), is essentially constant, indicating an overall improvement of about 20-25%, independent of the sample size.

As discussed, both the variance and the bias contribute to the MSE. Surprisingly, we found that — for this setup — the bias is almost negligible, with the squared bias contributing less than 1 % to the overall MSE at all n. Therefore, VI also yields a better variance than — despite what the name suggests — the minimum variance path by Blondel [104] and Pham and Shirts [105]. The reason is again that the latter was optimized under different assumptions and for a different estimator. However, note that while we have derived the sequence with the optimal MSE, the magnitude of the improvement compared to, e.g., linearly interpolated intermediates is system dependent, and, as we will see, actually can be much larger for many-particle systems than for one-dimensional ones.

3.5   Atomistic Test Cases

To compare the MSE of VI with that of established intermediates, we have performed test simulations for two atomistic many particle simulation systems, a Lennard-Jones gas and a solvated butanol molecule.

Lennard-Jones Gas

The free energy difference between an Argon and a Helium Lennard-Jones (LJ) gas with $M = 20$ atoms was calculated. A reference free energy difference was determined by conducting a long simulation with each method using 12 states with linearly spaced λ_s of both the linear intermediates and the approximated VI sequence and computation runs of 10 μs in each state. At this length, the relative difference of the estimates between the two methods is below 10^{-5} ($\Delta G = 0.23252 k_B T$). Using this reference value, the MSE of a distribution of 800 free energy differences determined with only five intermediates depending on the simulation time in each state was calculated.

In each state, the atoms were placed at random positions without overlap inside a cubic box. The atoms were assigned velocities drawn from the Boltzmann distribution corresponding to the temperature of $T = 298$ K. The simulations were conducted in the NVT ensemble at $T = 298$ K in a cubic box using periodic boundary conditions. The volume of the box was set to $(43.5 \text{ Å})^3$, corresponding to a pressure of about 10 bar. The atomic interaction at a distance r between the centers of two atoms was described through a Lennard-Jones potential

$$H(r) = 4\epsilon \left[\left(\frac{\sigma}{r} \right)^{12} - \left(\frac{\sigma}{r} \right)^{6} \right] \tag{3.30}$$

with parameters $\sigma = 3.405$ Å, $\epsilon = 1.0446$ kJ/mol and $m = 39.95$ u for Argon, and $\sigma = 2.64$ Å, $\epsilon = 0.0906$ kJ/mol and $m = 4$ u for Helium [237].

The leap-frog algorithm with a time step of 5 fs was used and velocity rescaling at every 20th time step. For both sequences, the 800 free energy simulations were carried out with 1 ns equilibration time and 5 ns production runs in each state. Five intermediate, i.e., seven states in total were used. In absence of further knowledge, equal spacing of λ_s and ζ_s, i.e, $\{0, 0.17, 0.33, 0.5, 0.67, 0.83, 1\}$ was used. For the approximated VI sequence (Eq.(3.29)), $C = 0$ was used throughout the

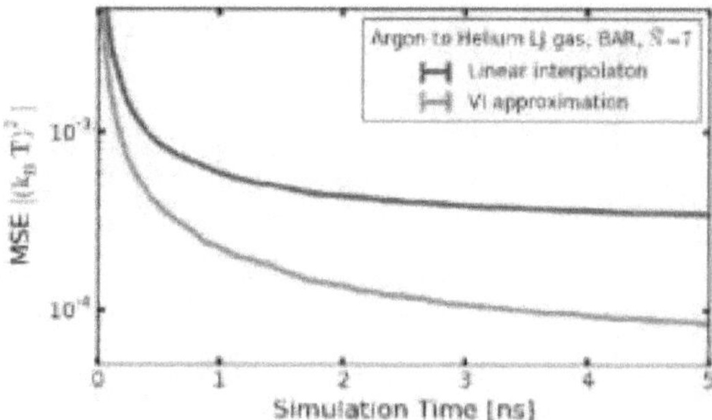

Figure 3.8: An Argon LJ gas is morphed into a Helium LJ gas. The MSE with respect to a converged reference value is shown as a function of the simulation time in each intermediate. It was obtained using linear intermediates (red) and the approximated VI intermediates (green). For both, an equal spacing of λ values, and ζ values, respectively, has been used.

whole simulation. The difference of the Hamiltonians between adjacent states was recorded at every 200th step. Free energy differences were subsequently calculated using BAR.

Figure 3.8 shows the resulting MSEs. For short simulation lengths, the MSE improves rapidly. For longer simulation times, the improvement of VI becomes most pronounced. At 5 ns, VI (green) has a four times lower MSE than conventional linear intermediates (red). Conversely, the MSE achieved by linear intermediates at 5 ns is already obtained at 0.56 ns by VI, which, at this level, thus requires almost 10 times less sampling.

Charge Decoupling of Butanol

For a last system closer to biomolecular applications, the approximated VI intermediates were implemented into GROMACS 2019. We calculated the solvation free energy difference between charged and uncharged butanol (15 atoms) solvated in water (1800 atoms in total). The topology from the SolvationToolkit package from Bannan et al. [222] was used. As for the Lennard-Jones gas, a reference value was obtained through extensive simulations, which then was compared to the estimates of a number of shorter simulations with fewer states. For the reference value we used 51 linear intermediates with equidistant λ states (i.e., $\Delta\lambda = 0.02$) and production runs of 100 ns simulation time in each state, totalling

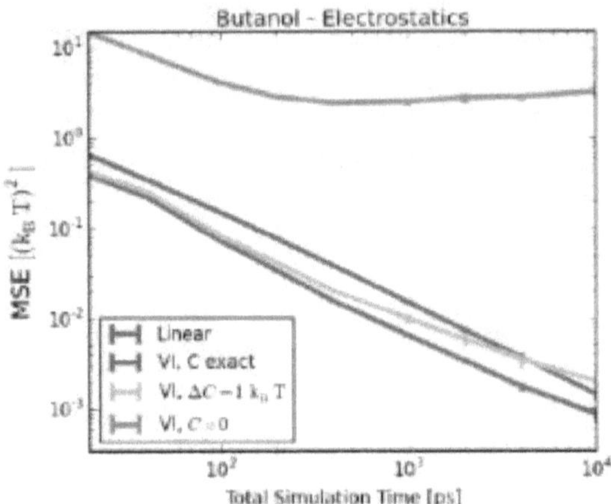

Figure 3.9: Accuracies of the estimates for the solvation free energy difference between charged and uncharged butanol in dependence of simulation time. Linear intermediates (red) are compared to VI using different initial guesses.

5.1 μs total simulation time. Energy values were recorded at every 200th step. Equilibration of 100 ps at constant volume and 200 ps at constant pressure with the Parrinello-Rahman barostat [228] was conducted prior to the production runs.

A reference value between the coupled and decoupled charges of 8.708 ± 0.001 $k_B T$ was obtained. Next, simulations in five different states were carried out with both conventional linear intermediates and VI using λ_s, and ζ_s values, respectively, of 0, 0.25, 0.5, 0.75 and 1. In each state, five production runs were conducted of 100 ns simulation time each, and smaller portions of the trajectories of different lengths were used to asses the MSEs as a function of trajectory length. For each trajectory length, a free energy difference estimate was calculated.

Figure 3.9 shows the obtained MSE for linear intermediates as well as for three different VI variants (see Supporting Information for a table of the MSEs), which differ in the choice of the initial guess: The MSE of the VI sequence with the exact estimate (blue) for C in Eq. (3.29) is about a factor of two better than the linear intermediates (red) at all simulation lengths, or equivalently, only half the amount of simulation time is necessary to obtain the same level of MSE. For an unrealistically large error in the initial guess C the MSE of the VI sequence is lower (green). However, with a more realistic initial estimate deviating by 1 k_BT from the reference value (yellow) the MSE also improved about a factor of two with respect to the linear intermediates, except for very long simulation lengths. Such estimates can, e.g., easily be obtained by linear intermediates with a simulation time of less than 40 ps — and therefore only a small fraction of the total simulation time — after which VI becomes the significantly better option than linear intermediates. In addition, the estimate can be further refined during the simulation process.

3.6 Conclusions

Using a variational principle, we have derived a minimum MSE sequence of intermediate Hamiltonians for free energy calculations using FEP and BAR. Our approach differs from previous ones in that it, firstly, optimizes the full MSE with respect to the exact free energy difference rather than the variance only (i.e., the precision). Secondly, it directly optimizes the sequence of discrete states, instead of a two step approach, where first a continuous TI path is optimized [78, 104, 105] and, subsequently, a discrete subset of states is chosen from this path. Thirdly, it holds for finite sampling and for any number of intermediate states, thus proving analytically that BAR is the optimal MSE estimator also for finite sampling.

We assessed the performance of our method using three test systems. First a simple one-dimensional model was considered. Compared to linear interpolations, a marked improvement in the MSEs was observed. Two limiting cases of our general VI result are notable. In the limit of many steps, the MSEs of the optimal VI and the MVP [78, 104, 105] are similar. In the limit of few steps, an approximated sequence was derived, the form of which appears similar to MVP, but differs in the exponent by factors of 2 and 1/2, respectively. However, for this model the smallest overall MSE was achieved for given computational effort at a medium number of intermediates (five in our case). Interestingly, the improvement in the MSE of the optimal VI compared to the MVP was mainly due to improvements of the variance; thus, the discretization of the path not only affects the bias, but also the variance.

Next, we considered an argon and a helium LJ gas and, somewhat closer to real applications for complex biomolecular systems, the solvation free energy difference between charged and uncharged butanol. For both many-particle test systems systems, marked improvements compared to conventional intermediates were seen.

This work focused on the theory and derivation of our variational approach. We have so far not tested our method on larger, more complex biomolecular systems involving conformational transitions. Therefore, further work is required towards the practical applicability, along several lines.

First, VI was derived assuming statistically independent sample points x_i. For atomistic simulation based sampling, as well as, to a lesser extent, for MC sampling, subsequent sample points are typically correlated, however, particularly when the relevant configuration space densities are separated by large barriers. In these cases, VI is not necessarily the optimal sequence, but can be combined with enhanced sampling techniques, such as Hamiltonian replica exchange [125, 128, 238], appropriate biasing potentials [110, 112, 239], or a combination thereof. Another possible route, indicated by the EDS method[100, 101], is to change the Hamiltonians such as to reduce energy barriers and, thereby, to reduce time-correlations. We are, however, unaware of any variational approach to optimize this trade-off and, therefore, further research will be required towards this aim.

Second, VI requires an initial estimate of the free energy differences. For all of our test cases, this requirement involved only little additional computational cost. Whether this remains true for more complex biomolecular systems, remains unclear at present.

Third, vanishing particles are a particular challenge due to possible singularities. Interestingly, preliminary simulations (data not shown) on systems with such vanishing LJ particles suggest that VI automatically generates intermediate Hamiltonians that resemble soft-core potentials. However, the singularities still caused instabilities in our test simulations. Smoothening the potential of the VI intermediates in these regions avoided this problem, suggesting that VI can also be used in this context. Clearly, additional work will be required to provide a widely applicable sequence for the disappearance of particles in solution.

3.7 Supporting Information

sim. time [ps]	linear	VI, C exact
20	$(6.46 \pm 0.04) \cdot 10^{-1}$	$(3.89 \pm 0.04) \cdot 10^{-1}$
40	$(3.50 \pm 0.03) \cdot 10^{-1}$	$(2.22 \pm 0.04) \cdot 10^{-1}$
100	$(1.49 \pm 0.02 \cdot 10^{-1}$	$(7.3 \pm 0.2) \cdot 10^{-2}$
200	$(7.7 \pm 0.2) \cdot 10^{-2}$	$(3.40 \pm 0.08) \cdot 10^{-2}$
400	$(3.8 \pm 0.1) \cdot 10^{-2}$	$(1.57 \pm 0.04) \cdot 10^{-2}$
1000	$(1.5 \pm 0.1) \cdot 10^{-2}$	$(6.6 \pm 0.3) \cdot 10^{-3}$
2000	$(7.5 \pm 0.5) \cdot 10^{-3}$	$(3.4 \pm 0.2) \cdot 10^{-3}$
4000	$(3.7 \pm 0.3) \cdot 10^{-3}$	$(1.8 \pm 0.2) \cdot 10^{-3}$
10000	$(1.4 \pm 0.1) \cdot 10^{-3}$	$(8 \pm 1) \cdot 10^{-4}$

Table 3.1: Mean squared errors (MSE) as a function of total simulation time for the electrostatic decoupling of solvated butanol. All MSEs are given in units of $[(k_B T)^2]$. The linear intermediates are compared to three variants of VI. Firstly, with an exact initial estimate of the free energy difference, secondly, with an estimate that is 1 $k_B T$ smaller than the exact reference value, and, thirdly, with an estimate of 0 $k_B T$, i.e., 8.708 $k_B T$ smaller than the reference value. For the latter two variants see continuation in Table 2.

sim. time [ps]	VI, $\Delta C = 1 k_B T$	VI, $C = 0$
20	$(4.2 \pm 0.1) \cdot 10^{-1}$	14.61 ± 0.06
40	$(2.6 \pm 0.1) \cdot 10^{-1}$	8.36 ± 0.06
100	$(8.2 \pm 0.4) \cdot 10^{-2}$	4.05 ± 0.05
200	$(4.1 \pm 0.2) \cdot 10^{-2}$	2.82 ± 0.05
400	$(2.0 \pm 0.1) \cdot 10^{-2}$	2.43 ± 0.06
1000	$(1.0 \pm 0.1) \cdot 10^{-2}$	2.50 ± 0.09
2000	$(5.9 \pm 0.9) \cdot 10^{-3}$	2.75 ± 0.14
4000	$(3.5 \pm 0.7) \cdot 10^{-3}$	2.82 ± 0.08
10000	$(1.9 \pm 0.6) \cdot 10^{-3}$	3.21 ± 0.21

Table 3.2: Continuation of Table 1

End of publication

3.8 Full MSE Result

All terms relevant to derive the optimal sequence of intermediate states have been calculated in section 3.4, whereas all terms irrelevant to the optimization have been dropped. However, to, e.g., predict MSEs for a given model system, the complete expression of the MSE is required. It is calculated in the following.

The quadratic terms have been extended in Eq. 3.16. Continuing by factorizing all integral terms yields

$$
\mathbb{E}\left[\left(\Delta G^{(n)}_{s' \to (s+1)'}\right)^2\right]
$$

$$
= \frac{1}{n^2} \sum_{i=1}^{n} \int p_s(\mathbf{x}_1)\, d\mathbf{x}_1 \cdots \int p_s(\mathbf{x}_n)\, d\mathbf{x}_n \left(e^{-(H'_{s+1}(\mathbf{x}_i)-H'_s(\mathbf{x}_i))} - 1\right)^2
$$

$$
+ \frac{1}{n^2} \sum_{\substack{i=1 \\ }}^{n} \sum_{\substack{j=1 \\ j \neq i}}^{n} \int p_s(\mathbf{x}_1)\, d\mathbf{x}_1 \cdots \int p_s(\mathbf{x}_n)\, d\mathbf{x}_n
$$

$$
\left(e^{-(H'_{s+1}(\mathbf{x}_i)-H'_s(\mathbf{x}_i))} - 1\right)\left(e^{-(H'_{s+1}(\mathbf{x}_j)-H'_s(\mathbf{x}_j))} - 1\right) \tag{3.31}
$$

$$
= \frac{1}{n} \int p_s(\mathbf{x})\, d\mathbf{x} \left(e^{-2(H'_{s+1}(\mathbf{x}_i)-H'_s(\mathbf{x}_i))} - 2e^{-(H'_{s+1}(\mathbf{x}_i)-H'_s(\mathbf{x}_i))} + 1\right)
$$

$$
+ \frac{n^2 - n}{n^2} \int p_s(\mathbf{x})\, d\mathbf{x} \left(e^{-(H'_{s+1}(\mathbf{x}_i)-H'_s(\mathbf{x}_i))} - 1\right)^2 \tag{3.32}
$$

$$
= \frac{1}{n}\left(\int p_s(\mathbf{x})\, d\mathbf{x}\, e^{-2(H'_{s+1}(\mathbf{x}_i)-H'_s(\mathbf{x}_i))} + 2\Delta G_{s' \to (s+1)'} - 1\right)
$$

$$
+ \left(1 - \frac{1}{n}\right)\left(\Delta G_{s' \to (s+1)'}\right)^2 \tag{3.33}
$$

$$
= \frac{1}{n}\left(\int p_s(\mathbf{x})\, d\mathbf{x}\, e^{-2(H'_{s+1}(\mathbf{x}_i)-H'_s(\mathbf{x}_i))} + \Delta G_{s' \to (s+1)'} - 1\right) + \left(\Delta G_{s' \to (s+1)'}\right)^2 . \tag{3.34}
$$

Similar expressions are obtained for $\mathbb{E}\left[\left(\Delta G^{(n)}_{s' \to (s+1)'}\right)^2\right]$.

Next, addressing the second line in Eq. 3.6, using Eq. 3.13, yields

$$-2\Delta G_{1,N} \left(\sum_{\substack{s=1 \\ s \text{ odd}}}^{N-2} \left(\mathbb{E}\left[\Delta G^{(n)}_{s \to s+1}\right] - \mathbb{E}\left[\Delta G^{(n)}_{s+2 \to s+1}\right] \right) \right) = -2(\Delta G_{1,N})^2 \qquad (3.35)$$

Similarly, the third line of Eq. 3.6 is treated as in Eq. 3.15. All terms without a prefactor of $1/n$, can, under the condition of Eq. 3.7, be summarized as a binomial and cancel with $(\Delta G_{1,N})^2$. As such, the remaining terms yield,

$$\begin{aligned}
\mathrm{MSE}&\left(\Delta G^{(n)}_{1,N}\right) \\
=\frac{1}{n}&\sum_{\substack{s=1 \\ s \text{ odd}}}^{N-2} \left(\int p_s(\mathbf{x})\,\mathrm{d}\mathbf{x}\, e^{-2(H'_{s+1}(\mathbf{x}_i)-H'_s(\mathbf{x}_i))} \right. \\
&+ \int p_{s+2}(\mathbf{x})\,\mathrm{d}\mathbf{x}\, e^{-2(H'_{s+1}(\mathbf{x}_i)-H'_{s+2}(\mathbf{x}_i))} \\
&\left. + \Delta G_{s' \to (s+1)'} + \Delta G_{(s+2)' \to (s+1)'} - 2 \right)
\end{aligned} \qquad (3.36)$$

where the last line corresponds to the constant expression g'_s in Eq. 3.18. Therefore, the MSE converges to zero with increasing n. However, as can be seen, the MSE is increased for non-zero free energy differences between the shifted Hamiltonians.

3.9 Solving the System of Equations

The system of Eqs. 3.22 and 3.23 defining the sequence of optimal Hamiltonians $\{H_s(\mathbf{x})\}$ also depends on the partition functions $\{Z_s\}$ of all states s, and *vice versa*. Whereas numerous numerical algorithms exist for solving systems of equations, this interdependence complicates the application to Eqs 3.22 and 3.23. For all one-dimensional model systems presented in this thesis, fixed point iteration (FPI) was used. The details of its application to VI are provided below. On the one hand side, FPI was stable in all cases and converged to a unique solution independent of the initial guess. On the other hand side, it is slow compared to most alternative algorithms and therefore the application of a faster stable algorithm would be desirable for the future.

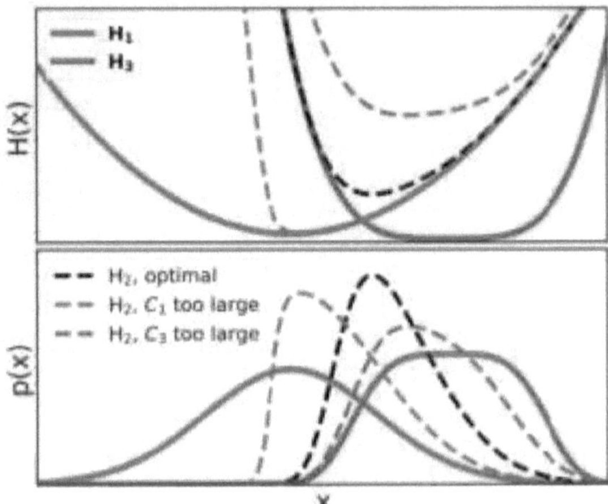

Figure 3.10: Different variants of a virtual intermediate state (dashed lines) between the end states (solid lines). The black dashed line depicts the optimal intermediate. The red and blue dashed lines show the intermediates, where one of the underlying shifting constants, C_1 and C_3, is too large. Top: Hamiltonian. Bottom: Configuration space density.

Influence of the Estimated Partition Functions

The form of an intermediate state s depends on the form of the adjacent states $s-1$ and $s+1$. For one virtual intermediate state between the two end states, the optimal intermediate, using 3.21, is defined via

$$e^{H_2(\mathbf{x})} = e^{H_1(\mathbf{x})-C_1} + e^{H_3(\mathbf{x})-C_3} \tag{3.37}$$

where the accuracy is optimal if the shifting constant $C_i = -\ln Z_i$ $(i = 1, 3)$. In this case,

$$p_2(\mathbf{x})^{-1} \propto p_1(\mathbf{x})^{-1} + p_3(\mathbf{x})^{-1} \ . \tag{3.38}$$

To illustrate how C_1 and C_3 influence the form of the intermediate, Figure 3.10 shows the Hamiltonian and configuration space density of different variants of a virtual intermediate (dashed lines) between the end states (solid lines). In the optimal case (black), $p_2(\mathbf{x})$ is largest in the regions where the configuration space densities of the end states overlap. If, however, either C_1 or C_3 (red and blue dashed line, respectively) are too large, then the form of the virtual intermediate becomes closer to either state 1 or 3. It has been tested that in this case the MSEs are suboptimal. Hence, the shifting constant involved in the following FPI largely influence the form of the intermediate states.

Fixed Point Iteration and VI

Denoting the set of all intermediate Hamiltonians as $\mathbf{H}(\mathbf{x}) = \{H_2(\mathbf{x}), ..., H_{N-1}(\mathbf{x})\}$, and the corresponding shifting constants as $\mathbf{C} = \{C_2, ..., C_{N-1}\}$, then the system of equations is solved through the iteration

$$\mathbf{H}^k(\mathbf{x}) = \mathbf{f}\left(\mathbf{H}^{k-1}(\mathbf{x}), \mathbf{C}^{k-1}\right) \tag{3.39}$$

$$\mathbf{C}^k = -\ln \int_{-\infty}^{\infty} e^{-\mathbf{H}^k(\mathbf{x})} d\mathbf{x} \ , \tag{3.40}$$

where k denotes the iteration step and

$$f\left(\mathbf{H}^k(\mathbf{x}), \mathbf{C}^k\right) = \begin{pmatrix} \ln\left[e^{H_1(\mathbf{x})-C_1} + e^{H_2^k(\mathbf{x})-C_2^k}\right] \\ -\frac{1}{2} \ln\left[e^{-2(H_2^k(\mathbf{x})-C_2^k)} + e^{-2(H_4^k(\mathbf{x})-C_4^k)}\right] \\ \ln\left[e^{H_3^k(\mathbf{x})-C_3^k} + e^{H_5^k(\mathbf{x})-C_5^k}\right] \\ -\frac{1}{2} \ln\left[e^{-2(H_4^k(\mathbf{x})-C_4^k)} + e^{-2(H_6^k(\mathbf{x})-C_6^k)}\right] \\ ... \\ \ln\left[e^{H_{N-1}^k(\mathbf{x})-C_{N-1}^k} + e^{H_N(\mathbf{x})-C_N}\right] \end{pmatrix} \ . \tag{3.41}$$

The integral notation in Eq. 3.40 denotes that at each step k and for each state s the integration

$$C_s^k = -\ln \int_{-\infty}^{\infty} e^{-H_s^k(\mathbf{x})} d\mathbf{x} \tag{3.42}$$

is conducted. As the end state Hamiltonians $H_1(\mathbf{x})$ and $H_N(\mathbf{x})$ are fixed, C_1 and C_N only need to be calculated once at the start of the FPI.

The integration, Eq. 3.40, is required at every step k due to the fact that $H_s^k(\mathbf{x})$ is calculated as the optimum between the adjacent intermediates $s-1$ and

$s + 1$ from the previous step $k - 1$. Once $\mathbf{C}^k$ has been determined up to a step κ for $k = 0, ..., \kappa$, the optimal set of Hamiltonians $\mathbf{H}^\kappa(\mathbf{x})$ of a configuration $\mathbf{x}$ that had previously not been considered, is determined through iteration of Eq. 3.39 only. Note that, to be able to do so, it is necessary to store $\mathbf{C}^k$ in memory for all steps, such that the iteration of Eq. 3.39 can be conducted.

The FPI is completed once a convergence criterion is fulfilled. For the cases in this chapter, the criterion

$$\left| \frac{C_s^k - C_s^{k-1}}{C_s^k} \right| \leq 10^{-6} \quad \forall s \tag{3.43}$$

was used.

As for the initial guess $\mathbf{H}^0(\mathbf{x})$: Whereas it was tested that the FPI reliably converges to a unique solution independent of the initial guess, convergence is achieved in fewer steps if $\mathbf{H}^0(\mathbf{x})$ is closer to the optimal form. Therefore, the approximated VI form derived in section 3.5, which is relatively close to the exact VI form, was used, i.e.,

$$\mathbf{H}^0(\mathbf{x}) = \begin{pmatrix} -\frac{1}{2} & \ln\left[(1 - \lambda_2)e^{-2(H_A(\mathbf{x})-C_A)} + \lambda_2 e^{-2(H_B(\mathbf{x})-C_B)}\right] \\ -\frac{1}{2} & \ln\left[(1 - \lambda_3)e^{-2(H_A(\mathbf{x})-C_A)} + \lambda_3 e^{-2(H_B(\mathbf{x})-C_B)}\right] \\ & \cdots \\ -\frac{1}{2} & \ln\left[(1 - \lambda_{N-1})e^{-2(H_A(\mathbf{x})-C_A)} + \lambda_{N-1} e^{-2(H_B(\mathbf{x})-C_B)}\right] \end{pmatrix} . \tag{3.44}$$

For all test systems, the $\{\lambda_s\}$ were chosen as $\lambda_s = (s - 1)/(N - 1)$, i.e., in equidistant steps.

For high-dimensional many-body systems, performing multiple integrations over the entire configuration space is, by nature of the problem, infeasible. However, it is possible to perform the first few steps of the FPI: Firstly, a set of simulations is conducted with the approximated VI sequence, i.e., the initial guess. Based on estimates of the free energy differences between adjacent states, the relative shift constants $C_s^0 - C_1$ are calculated for all states s. In the next step, simulations are conducted in states governed by $\mathbf{H}_s^1$ determined by Eq. 3.39, and so on. Thereby, the resulting form of the intermediate states will become closer to the optimal one than the form of the approximated VI sequence.

3.10   Exponential Error Metrics

The VI derivation is based on two assumptions:

1. Sample points are independent.

2. A set of shifting constants $\{C_s\}$ exists such that both $\Delta G_{s\to t} \ll 1\ k_BT$ and $\Delta G^{(n)}_{s\to t} \ll 1\ k_BT$ for $t = s - 1$ and $t = s + 1$. Therefore, as indicated in section 3.5, even the exact VI form may not be optimal in the rare case of very few samples point and very small overlap in configuration space density between adjacent states.

This sections outlines how, in theory, the violation of these two assumptions can be distinguished for cases with suboptimal MSEs.

The following error metric, referred to as the Exponential Mean Squared Error (EMSE), is defined as

$$
\text{EMSE}\left(\Delta G^{(n)}_{1,N}\right)
$$

$$
= \mathbb{E}\left[\left(\sum_{\substack{s=1\\ s\ \text{odd}}}^{N-3} -e^{-\Delta G_{s\to s+1}} + e^{-\Delta G_{s+2\to s+1}} + e^{-\Delta G^{(n)}_{s\to s+1}} - e^{-\Delta G^{(n)}_{s+2\to s+1}}\right)^2\right].
$$

$$(3.45)$$

Upon insertion of all $\Delta G_{s\to t}$ and $\Delta G^{(n)}_{s\to t}$ ($t = s - 1$ and $t = s + 1$) into Eq. (3.45), the exponentials cancel with the logarithms in the definitions of the free energy differences. The subsequent expression is identical to Eq. 3.18, which was minimized in section 3.5 and yielded the VI.

Importantly, the VI are therefore obtained by variation of the EMSE without any analytical approximation. As a consequence, for independent sampling, the EMSE of the VI sequence is always optimal, also in cases where the MSE is not. Furthermore, if $\Delta G^{(n)}_{s\to t}$ and $\Delta G_{s\to t}$ are small compared to k_BT, then $\text{EMSE}\left(\Delta G^{(n)}_{1,N}\right) \approx \text{MSE}\left(\Delta G^{(n)}_{1,N}\right)$. Therefore, in case assumption 2 is violated, the MSE and EMSE differ. Alternatively, if the samples are not independent (i.e., violation of assumption 1), then both the MSE and EMSE would be affected, and the EMSE of VI would therefore also be suboptimal.

For test simulations, one further change is required: The EMSE is, unlike the MSE, not invariant against constant shifts in the Hamiltonians, such as

$H'_s(\mathbf{x}) = H_s(\mathbf{x}) - C_s$. Therefore, the EMSE can simply be minimized by choosing a constant such that $e^{-\Delta G_{s \to s+1}}$ is small, which would, however, not enable any conclusions about the underlying form with respect to the MSE. Therefore, one option to identify optimal forms in test simulations is to compare the EMSE between states with equal partition functions only (i.e., similar to the analytical derivation, where the partition sums of the intermediates are restrained to a constant through Lagrange multipliers). Alternatively, the Relative Exponential Mean Squared Error (REMSE) metric,

$$\mathrm{REMSE}\left(\Delta G_{1,N}^{(n)}\right)$$

$$= \mathbb{E}\left[\left(\sum_{\substack{s=1 \\ s \text{ odd}}}^{N-2} \frac{-e^{-\Delta G_{s \to s+1}} + e^{-\Delta G_{s \to s+1}^{(n)}}}{e^{-\Delta G_{s \to s+1}}} + \frac{e^{-\Delta G_{s+2 \to s+1}} - e^{-\Delta G_{s+2 \to s+1}^{(n)}}}{e^{-\Delta G_{s+2 \to s+1}}}\right)^2\right],$$

$$(3.46)$$

may be used. Here, the difference in each individual step is normalized by the exponentially weighted difference itself. As can be easily validated, $\mathrm{REMSE}\left(\Delta G_{1,N}^{(n)}\right)$ is invariant to any constant shifts in the intermediate Hamiltonians. For small free energy differences $\Delta G_{s \to t}^{(n)}$ and $\Delta G_{s \to t}$, the REMSE reduces to the MSE, as desired.

4

Correlated Free Energy Estimates

4.1 Variationally derived intermediates for correlated free-energy estimates between intermediate states

The following section consists of the article

M. Reinhardt, H. Grubmüller, "Variationally derived intermediates for correlated free-energy estimates between intermediate states", *Physical Review E*, vol. 102, issue 4, p. 043312 (2020) [240]

published under the Creative Commons Attribution 4.0 International license and the content reprinted based on the 2020 Copyright Agreement by the American Physical Society with consent of the authors.

The format, including the numbering of equations and figures, has been altered to match the format of this thesis. Furthermore, all references are listed in the bibliography at the end of this thesis rather than at the end of this article.

Both authors contributed to conceiving the study and writing the manuscript. I implemented, conducted and analyzed all test simulations.

4.2 Introduction

Free energy calculations are widely used to investigate physical and chemical processes [1–7]. Their accuracy is essential to biomedical applications such as computational drug development [8–11] or material design [12–15]. Amongst the most widely used methods based on simulations with atomistic Hamiltonians are alchemical equilibrium techniques, including the Free Energy Perturbation (FEP) [79] and Thermodynamic Integration (TI) [90] methods. These techniques determine the free energy difference between two states, representing, for example, two different ligands bound to a target, by sampling from intermediate states whose Hamiltonians are constructed from those of the end states. The free energy difference between the end states is then determined via a step-wise summation of the differences between the intermediate states.

The choice of these intermediates critically affects the accuracy of the free energy estimates [165, 231, 241] by determining which parts of the configuration space are sampled to which extent [169], thereby performing a function similar to importance sampling [78]. In addition, different estimators that determine the free energy differences between these intermediates and the end states have been developed, most prominently the Zwanzig formula [79] for FEP, the Bennett Acceptance Ratio method (BAR) [80], and multistate BAR (MBAR) [166].

We have recently derived [230] the sequence of discrete intermediate states—the variationally derived intermediates (VI)—that yield, for finite sampling, the lowest mean squared error (MSE) of the free energy estimates with respect to the exact value. Their form differs from the most common scheme, which, for N states, linearly interpolates between the end states Hamiltonians $H_1(\mathbf{x})$ and $H_N(\mathbf{x})$, respectively, along a path variable λ_s.

$$H_s(\mathbf{x}) = (1 - \lambda_s)H_1(\mathbf{x}, \lambda_s) + \lambda_s H_N(\mathbf{x}, \lambda_s), \quad \lambda_s \in [0,1] \tag{4.1}$$

where $\mathbf{x} \in \mathbb{R}^{3M}$ denotes the coordinate vector of all M particles in the system. All states are labeled by an integer s with $1 \leq s \leq N$, and λ_s corresponding to state s. The additional λ_s argument of the end states Hamiltonians indicates the common use of soft-core potentials [91–93] to avoid divergences for vanishing particles. Other approaches involve the interpolation of exponentially weighted Hamiltonians of the end states, such as Enveloping Distribution Sampling [100] (and variants thereof [102, 103]) or the Minimum Variance path [104, 105] for TI.

In contrast, the VI are not directly defined via the end states; instead, the optimal form of each intermediate s is determined by the form of the adjacent ones $s-1$ and $s+1$. For the setup shown in Fig. 4.1(a), which consists of two types of intermediates, sampling is conducted in the first type labeled with even numbered s and indicated by the solid lines with yellow points. These are governed by the optimal Hamiltonian

$$H_s(\mathbf{x}) = -\frac{1}{2} \ln[e^{-2H_{s-1}(\mathbf{x})} \cdot r_{s-1,s}^{-2} + e^{-2H_{s+1}(\mathbf{x})} \cdot r_{s+1,s}^{-2}], \tag{4.2}$$

where $r_{s,t} = Z_s/Z_t$ denotes the ratio of the configurational partition sums of states s and t. Virtual intermediates are the second type, and labeled with odd numbers s with $2 < s < N-1$ and indicated by the dashed lines in Fig. 4.1(a). For these,

$$H_s(\mathbf{x}) = \ln[e^{H_{s-1}(\mathbf{x})} \cdot r_{s-1,s} + e^{H_{s+1}(\mathbf{x})} \cdot r_{s+1,s}]. \tag{4.3}$$

Virtual intermediates are used as target states to evaluate the difference in free energy to, and no sampling is conducted in those. Due to the coupling of the VI, the optimal MSE sequence of Hamiltonians $H_2(\mathbf{x})...H_{N-1}(\mathbf{x})$ is determined by solving the system of $N-2$ equations of Eqs. (4.2) and (4.3).

The variational MSE minimization has been conducted based on the Zwanzig formula [79]

$$\Delta G_{s,s+1} = -\ln\langle e^{-[H_{s+1}(\mathbf{x})-H_s(\mathbf{x})]}\rangle_s \tag{4.4}$$

being used to calculate the difference between two adjacent states, as indicated

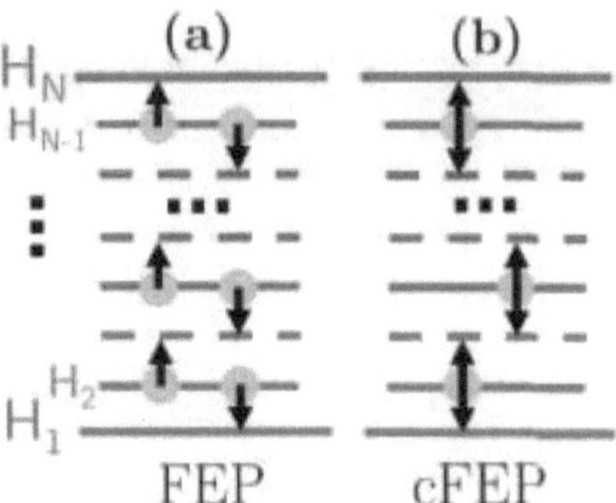

Figure 4.1: Two schemes of free energy calculation. The arrows indicate the Zwanzig formula is used to evaluate the free energy difference to the adjacent state based on sample sets represented through yellow dots. The dashed lines represent virtual intermediate states that no sampling is conducted in. (a) Separate and uncorrelated sample set are used to calculate the free energy difference of the respective intermediate to the state above and below (b). The same sample set is used for this purpose.

by the arrows in Fig. 4.1. However, using the virtual target states described by Eq. (4.3) is equivalent to using BAR directly between two sampling states [186, 230], and, therefore, Eq. (4.3) also describes the optimal intermediates for BAR.

Note that the Hamiltonians of the optimal intermediates, Eqs. (4.2) and (4.3), depend on the ratios of the partition sums, i.e., the desired quantity. Therefore, the system of equations has to be solved iteratively. The principle is the same as for BAR, where the estimator depends on an estimate of the free energy difference, and the optimal estimator is, therefore, determined iteratively. However, BAR is, in practice, mostly used with sampling states that are governed by Eq. (4.1) and a user chosen λ value. The VI method generalizes the BAR principle and determines not only the estimator, but the form of all intermediates through such an iterative optimization by using the information of the free energy estimates between these states. The resulting form differs from Eq. (4.1) and does not require any additional user choice of a λ variable.

However, for both BAR and VI to be optimal for multiple states, the free energy estimates to the states above and below an intermediate in the sequence have to be based on separate, uncorrelated sample points. This is illustrated by the separate yellow points in Fig. 4.1(a) that we refer to as the regular FEP setup, which was the topic of our previous work [230]. Yet, it would be twice as efficient to use the same sample points in both directions, as illustrated by Fig. 4.1(b), and as generally done in practice. However, this introduces correlations between the estimates to both adjacent intermediates, thereby violating the assumptions underlying the derivation of Eqs. (4.2) and (4.3). Therefore, in this case, BAR and the above variational intermediates are not optimal anymore. Due to these correlations, we refer to the Fig. 4.1(b) as the correlated FEP (cFEP) setup, which is the topic of this work.

Here, we derive the minimal MSE sequence of intermediate states and the corresponding estimators for cFEP, as used in practice, that take these correlations properly into account. This is in contrast to the derivation of VI [230] for FEP, where these correlations do not occur. As will be shown below, what might seem as a minor technical twist, markedly changes the shape of the optimal intermediates and considerably improves the accuracy of the obtained free energy estimates.

4.3 Theory

For the cFEP scheme shown in Fig. 4.1(b), we aim to derive the sequence of intermediate Hamiltonians $H_2(\mathbf{x}) \ldots H_{N-1}(\mathbf{x})$ that optimizes the MSE

$$\mathrm{MSE}\left(\Delta G^{(n)}\right) = \mathbb{E}\left[\left(\Delta G - \Delta G^{(n)}\right)^2\right] \tag{4.5}$$

along similar lines as for the regular FEP scheme [230], shown in Fig. 4.1(a). Here, $\Delta G_{1,N}^{(n)}$ denotes the free energy estimate based on a finite number of sample points n, and $\Delta G_{1,N}$ the exact difference between the end states 1 and N.

For the optimization metrics, different choices are possible, such as the Kullback-Leibler divergence [242] or the Fisher information metric [165, 243] that measure the (dis)similarity between configuration space densities. Instead, here we chose to directly optimize the MSE, as it quantifies the average accuracy with respect to the exact free energy difference, which is the relevant measure for most practical applications. Furthermore, as the MSE can be decomposed into variance plus bias squared, we account for both of these contributions that are oftentimes optimized separately in the literature [171, 234].

The cFEP variant in Fig. 4.1(b) only uses sampling in the intermediate states. Setups that, in addition, involve sampling in the end states, can also be treated with the formalism below. However, firstly, as we have tested, the accuracy for a given computational effort does not increase in this case. Secondly, mixing two different types of sample points (the ones used to evaluate ΔH to only one adjacent state vs. to both adjacent states) further complicates the analysis.

For cFEP, the estimated difference is

$$\Delta G^{(n)} = \sum_{\substack{s=2 \\ s \text{ even}}}^{N-2} \left(\Delta G^{(n)}_{s \to s+1} - \Delta G^{(n)}_{s \to s-1} \right) . \tag{4.6}$$

As in Fig. 4.1(b), the arrows point from sampling to target states, i.e., either the end states or the virtual intermediates. Assuming for each sample state s a set of n independent sample points $\{\mathbf{x}_i\}$, drawn from $p_s(\mathbf{x}) = e^{-H_s(\mathbf{x})}/Z_s$, with partition function Z_s, expanding Eq. (4.5) with the use of Eq. (4.6) reads

$$\mathrm{MSE}\left(\Delta G^{(n)}_{1,N} \right) =$$

$$(\Delta G_{1,N})^2 + \sum_{\substack{s=2 \\ s \text{ even}}}^{N-2} \mathbb{E}\left[\left(\Delta G^{(n)}_{s \to s+1} \right)^2 + \left(\Delta G^{(n)}_{s \to s-1} \right)^2 \right]$$

$$- 2\Delta G_{1,N} \left(\sum_{\substack{s=2 \\ s \text{ even}}}^{N-2} \left(\mathbb{E}\left[\Delta G^{(n)}_{s \to s+1} \right] - \mathbb{E}\left[\Delta G^{(n)}_{s \to s-1} \right] \right) \right) \tag{4.7}$$

$$- \sum_{\substack{s=2 \\ s \text{ even}}}^{N-2} \sum_{\substack{t=2 \\ t \text{ even}}}^{N-2} \mathbb{E}\left[2\,\Delta G^{(n)}_{s \to s+1}\, \Delta G^{(n)}_{t \to t-1} \right] .$$

The first two lines of Eq. (4.7) have already been processed in Ref. 230, but the last term differs. Previously, as in the regular FEP scheme in Fig. 4.1(a), these last expectation values were originally derived from independent sample sets and were, therefore, uncorrelated. In the present context of cFEP, however, these estimates are correlated. Therefore, the term needs to be split in two sums, distinguishing between the pairs with samples from the same state and the ones from different states,

$$\sum_{\substack{s=2\\ s\,\text{even}}}^{N-2} \sum_{\substack{t=2\\ t\,\text{even}}}^{N-2} E\left[2\,\Delta G^{(n)}_{s\to s+1}\,\Delta G^{(n)}_{t\to t-1}\right]$$

$$= 2\sum_{\substack{s=2\\ s\,\text{even}}}^{N-2} E\left[\Delta G^{(n)}_{s\to s+1}\,\Delta G^{(n)}_{s\to s-1}\right] \tag{4.8}$$

$$+ 2\sum_{\substack{s=2\\ s\,\text{even}}}^{N-2}\sum_{\substack{t=2\\ t\,\text{even}\\ t\neq s}}^{N-2} E\left[\Delta G^{(n)}_{s\to s+1}\right] E\left[\Delta G^{(n)}_{t\to t-1}\right].$$

where the expectation value of the product between the two estimates based on different sample sets has been separated, as these are uncorrelated.

As we are only interested in the intermediates that optimize the MSE, and not in the absolute value of the MSE, we focus on the terms that will not drop out in the optimization below.

Continuing with the expression inside the sum of the first term on the right hand side of Eq. (4.8),

$$E\left[\Delta G^{(n)}_{s\to s+1}\,\Delta G^{(n)}_{s\to s-1}\right] \tag{4.9}$$

$$= -\int p_s(\mathbf{x}_1)\mathrm{d}\mathbf{x}_1 \cdots \int p_s(\mathbf{x}_n)\mathrm{d}\mathbf{x}_n$$

$$\ln\left[\frac{1}{n}\sum_{i=1}^n e^{-(H_{s+1}(\mathbf{x}_i)-H_s(\mathbf{x}_i))}\right] \tag{4.10}$$

$$\ln\left[\frac{1}{n}\sum_{i=1}^n e^{-(H_{s-1}(\mathbf{x}_i)-H_s(\mathbf{x}_i))}\right].$$

As in the derivation of Ref. 230, the Hamiltonians are now shifted by a constant offset C_s, i.e., $H'_s(\mathbf{x}) = H_s(\mathbf{x}) - C_s$. This offset will cancel out for a given shape of an intermediate when calculating the accumulated free energy difference in Eq. (4.6). However, as the intermediate states will turn out to be coupled, these offsets do influence the shape of these intermediates. The offsets can now be chosen such that the terms inside the logarithms of Eq. (4.10) are close to one. In this case, $E\left[\Delta G^{(n)}_{s'\to(s+1)'}\right] = \Delta G_{s',(s+1)'}$ [230], and, therefore, the two linear terms arising from Eq. (4.10) can be expressed in terms of the exact free energy differences.

Next, the product of the two sums in Eq. (4.10) is split into terms based on the same and different sample points, respectively.

$$\mathbb{E}\left[\Delta G^{(n)}_{s'\to(s+1)'}\,\Delta G^{(n)}_{s'\to(s-1)'}\right] \tag{4.11}$$

$$= -\frac{1}{n^2}\int p_s(\mathbf{x}_1)\mathrm{d}\mathbf{x}_1\dots\int p_s(\mathbf{x}_n)\mathrm{d}\mathbf{x}_n$$

$$\left[\left(\sum_{i=1}^{n}e^{-(H'_{s+1}(\mathbf{x}_i)-H'_s(\mathbf{x}_i))}\right)\left(\sum_{\substack{j=1\\j\neq i}}^{n}e^{-(H'_{s-1}(\mathbf{x}_j)-H'_s(\mathbf{x}_j))}\right)\right. \tag{4.12}$$

$$\left.+\sum_{i=1}^{n}e^{-H'_{s+1}(\mathbf{x}_i)-H'_{s-1}(\mathbf{x}_i)+2H'_s(\mathbf{x}_i)}\right]$$

$$+f_{s'}\left(\Delta G_{s'\to(s-1)'},\Delta G_{s'\to(s+1)'}\right),$$

where the terms that can be expressed solely based on (constant) free energy differences are summarized by the term f_s. Again, the first two terms of Eq. (4.12) can be expressed in terms of the free energy differences between s and $s+1$ as well as between s and $s-1$, respectively.

Collecting all terms arising from Eq. (4.7)

$$\mathrm{MSE}\left(\Delta G^{(n)}_{1,N}\right)$$

$$=\sum_{\substack{s=2\\s\text{ odd}}}^{N-2}\frac{1}{n}\left(\int p_s(\mathbf{x})\,\mathrm{d}\mathbf{x}\,e^{-2(H'_{s+1}(\mathbf{x})-H'_s(\mathbf{x}))}\right.$$

$$+\int p_{s+2}(\mathbf{x})\,\mathrm{d}\mathbf{x}\,e^{-2(H'_{s+1}(\mathbf{x})-H'_{s+2}(\mathbf{x}))} \tag{4.13}$$

$$+\int p_{s+1}(\mathbf{x})\,\mathrm{d}\mathbf{x}\,e^{-H'_{s+2}(\mathbf{x})-H'_s(\mathbf{x})+2H_{s+1}(\mathbf{x})}$$

$$\left.+g_{s'}\left(\Delta G_{s',(s+1)'},\Delta G_{(s+2)',(s+1)'},\Delta G_{1',N'}\right)\right),$$

where the function g'_s serves the same purpose as f'_s and can be dropped in the optimization below.

The condition of small $\Delta G^{(n)}_{s'\to(s+1)'}$ is fulfilled by setting $C_s = -\ln Z_s$. By variation of the MSE from Eq. (4.13),

$$\frac{\partial}{\partial H_s(\mathbf{x})}\left(\mathrm{MSE}\left(\Delta G^{(n)}_{1,N}\right)+\nu\int(e^{-H_s(\mathbf{x})}-Z_s)\mathrm{d}\mathbf{x}\right)\overset{!}{=}0, \tag{4.14}$$

where ν is a Lagrange multiplier, the optimal sequence of Hamiltonians is obtained. For s *even*, we obtain

$$H_s(\mathbf{x}) = -\frac{1}{2} \ln \left(e^{-2H_{s-1}(\mathbf{x})} r_{s-1,s}^{-2} + e^{-2H_{s+1}(\mathbf{x})} r_{s+1,s}^{-2} \right.$$
$$\left. -2 e^{-H_{s-1}(\mathbf{x}) - H_{s+1}(\mathbf{x})} r_{s-1,s}^{-1} r_{s+1,s}^{-1} \right) \tag{4.15}$$

For s *odd* and $2 < s < N - 1$:

$$H_s(\mathbf{x}) = \ln \left(e^{H_{s-1}(\mathbf{x})} r_{s-1,s} + e^{H_{s+1}(\mathbf{x})} r_{s+1,s} \right)$$
$$- \ln \left(e^{-H_{s-2}(\mathbf{x}) + H_{s-1}(\mathbf{x})} r_{s-1,s-2} \right. \tag{4.16}$$
$$\left. + e^{-H_{s+2}(\mathbf{x}) + H_{s+1}(\mathbf{x})} r_{s+1,s+2} \right)$$

where, as in Eqs. (4.2) and (4.3), the ratios $r_{s,t}$ of the partition sums between states s and t have to be determined iteratively. The above sequence, Eqs. (4.15) and (4.16), that we refer to as the correlated Variational Intermediates (cVI), yield the minimal MSE estimates for cFEP.

Figure 4.2 shows the resulting configuration space densities of the above intermediates for the example of a start state with a harmonic Hamiltonian, $H_1(\mathbf{x}) = \frac{1}{2}x^2$, and an end state with a quartic one, $H_N(\mathbf{x}) = (x - x_0)^4$. Panel (a) shows the VI that are optimal for the regular FEP scheme in Fig. 4.1(a). Panel (b) shows the cVI, optimal for cFEP.

The yellow (light) areas in Fig. 4.2, Eq. (4.17), provide a simple measure of the configuration space density overlap K between the end states 1 and N.

$$K = \int_{-\infty}^{+\infty} d\mathbf{x} \, \min(p_A(\mathbf{x}), p_B(\mathbf{x})), \tag{4.17}$$

Here, $K = 0$ indicates two separate distributions without any overlap, and $K = 1$ full overlap, i.e., identical configuration space densities.

The two rows in Fig. 4.2(a) and (b) depict the result for two different values of x_0, and correspondingly, varying K.

As can be inferred from Eq. (4.15), for $N = 3$, $H_2(\mathbf{x})$ diverges at the points where $p_1(\mathbf{x}) = p_3(\mathbf{x})$, and therefore, $p_2(\mathbf{x}) = 0$ at these points, as can also be seen for the intermediate sampling state shown in Fig. 4.2(a). More generally, $H_2(\mathbf{x})$ of cVI "directs" sampling away from the overlap regions and towards the ones

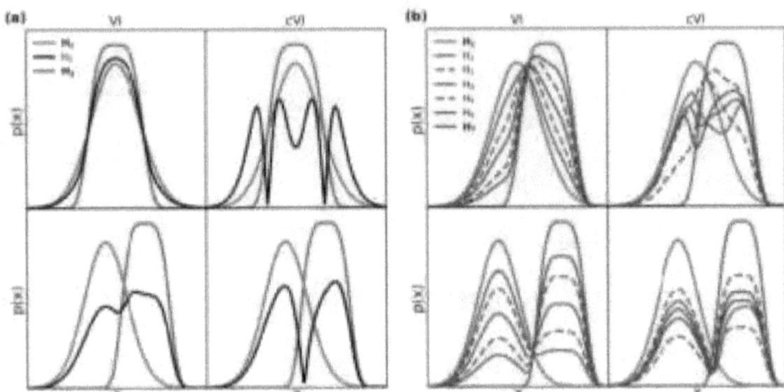

Figure 4.2: Configuration space densities of VI (left column), and cVI (right column) for (a) $N = 3$ and (b) $N = 7$ states. The individual rows show different shifts in x-direction between the minima of the harmonic, $H_1(\mathbf{x})$ (red, towards the left in each panel), and the quartic, $H_N(\mathbf{x})$ (blue, towards the right), potentials of the end states, thereby showing setups with different configuration space density overlap K between the end states, indicated by the yellow (light) area. Sampling is conducted in the even numbered intermediates. The dashed lines in (b) indicate the (odd numbered) virtual intermediate target states that no sampling is conducted in.

that are only relevant for one, but not both end states. For instance, the tails of the start state in the upper row of (a) are sampled more for cVI than for VI. For larger horizontal shifts of x_0, i.e., low values of K, the two variants become increasingly similar, as the additional term in Eq. (4.15) with respect to Eq. (4.2) becomes smaller compared to the first term.

For $N = 7$ states, Fig. 4.2(b) shows the converged resulting configuration space densities. The case of $x_0 = 0$, as shown in (a), was omitted in (b) as the visualization is more difficult in this case due to the higher number of states. In (b), the additional changes from VI to cVI become more complex. As in (a), the sampling states have smaller densities $p(\mathbf{x})$ in the overlap regions of the end states, but, in contrast to (a), still differ between VI and cVI for smaller values of overlap K. The reason is that while the overlap between the end states vanishes with decreasing K, an overlap between adjacent intermediate states remains that affects the shape of the intermediates. Note that the divergences mentioned above introduce instabilities in solving the system of Eqs. (4.15) and (4.16). Hence, for $N > 3$ the factor 2 of the additional term in the logarithm Eq. (4.15) has been replaced by a factor κ that was set to slightly below 2 ($\kappa = 1.95$) in case of Fig. 4.2(b). See Appendix A for details.

cBAR Estimator

As mentioned above, using the Zwanzig formula [79] to evaluate the free energy difference between two sampling states with respect to the virtual intermediate, Eq. (4.3), of VI is equivalent to BAR [186, 230]. Correspondingly, the virtual intermediate defined by Eq. (4.16) of cVI also corresponds to an estimator, that is optimal for the sampling states of cFEP and that we will refer to as correlated BAR (cBAR).

To derive cBAR, we use the relation between the two approaches. Determining the free energy difference between two sampling states labeled $s-1$ and $s+1$ by using the virtual intermediate s to evaluate the difference between the adjacent states yields

$$\Delta G^{(n)}_{s-1,s+1} = -\ln \frac{\langle e^{-(H_s(\mathbf{x}) - H_{s+1}(\mathbf{x}))}\rangle_{s+1}}{\langle e^{-(H_s(\mathbf{x}) - H_{s-1}(\mathbf{x}))}\rangle_{s-1}}. \tag{4.18}$$

Using the approach of Bennett [80] instead,

$$\begin{aligned}
&\Delta G^{(n)}_{s-1,s+1} \\
&= \ln \frac{\langle w(H_{s-1}(\mathbf{x}), H_{s+1}(\mathbf{x})) e^{-H_{s-1}(\mathbf{x})}\rangle_{s+1}}{\langle w(H_{s-1}(\mathbf{x}), H_{s+1}(\mathbf{x})) e^{-H_{s+1}(\mathbf{x})}\rangle_{s-1}}.
\end{aligned} \tag{4.19}$$

where $w(H_{s-1}(\mathbf{x}), H_{s+1}(\mathbf{x}))$ is a weighting function. From Eqs. (4.21) and (4.19) follows that the two approaches are equivalent if the weighting function relates to the Hamiltonian of the virtual intermediate state through

$$w(H_{s-1}(\mathbf{x}), H_{s+1}(\mathbf{x})) = e^{-H_s(\mathbf{x}) + H_{s-1}(\mathbf{x}) + H_{s+1}(\mathbf{x})}. \tag{4.20}$$

Therefore, any Hamiltonian of a virtual intermediate state corresponds to a weighting function. Bennett optimized the weighting function with respect to the variance yielding the famous BAR result

$$\Delta G^{(n)}_{s-1,s+1} - C = \ln \frac{\langle f(H_{s-1}(\mathbf{x}) - H_{s+1}(\mathbf{x}) - C)\rangle_{s+1}}{\langle f(H_{s+1}(\mathbf{x}) - H_{s-1}(\mathbf{x}) + C)\rangle_{s-1}}, \tag{4.21}$$

where $C \approx \Delta G_{s-1,s+1}$ has to be determined iteratively and $f(x)$ is the Fermi function. This result is equivalent to using the virtual intermediate of Eq. (4.3) with Eq. (4.18). Note that the relation of a virtual intermediate to BAR result had already been obtained by Lu et al. [186], albeit through a different formalism, and that using the hyperbolic secant function (Eq. 10, p. 2980), in their Overlap Sampling approach [186, 187] is equivalent to Eq. (4.20).

Next, for cFEP, using the Hamiltonian of the virtual intermediate from Eq. (4.16) in Eq. (4.20) yields the weighting function of cBAR,

$$
\begin{aligned}
w\big(&H_{s-2}(\mathbf{x}), H_{s-1}(\mathbf{x}), H_{s+1}(\mathbf{x}), H_{s+2}(\mathbf{x}), \\
&C_{s-2,s-1}, C_{s-1,s+1}, C_{s+1,s+2}\big) \\
=&\Big(e^{-H_{s-2}(\mathbf{x})+H_{s-1}(\mathbf{x})+C_{s-2,s-1}} \\
&\quad e^{-H_{s+2}(\mathbf{x})+H_{s+1}(\mathbf{x})+C_{s+2,s+1}}\Big)\Big/ \\
&\Big(e^{H_{s-1}(\mathbf{x})-H_{s+1}(\mathbf{x})-C_{s+1,s-1}} + 1\Big) ,
\end{aligned}
\tag{4.22}
$$

where the MSE of the resulting estimates is minimal if all $C_{s,t} \approx \Delta G_{s,t}$. A numerator of 1 in Eq. (4.22) would yield the original BAR result.

Note that $H_{s-2}(\mathbf{x})$, and $H_{s+2}(\mathbf{x})$, are also virtual intermediates determined by Eq. (4.16). As such, the result is a system of weighting functions, i.e., one for every pair of adjacent sampling states. The optimal estimate can, therefore, only be found by iteratively solving for the free energy estimates between all sampling states at once. In this regard, the procedure is similar to MBAR[166].

4.4 Test Simulations

To assess to what extent our new variational scheme improves accuracy, we consider the one-dimensional system with a harmonic and a quartic end state shown in Fig. 4.2. Rejection sampling is used to obtain uncorrelated sample points. The free energy estimate, obtained from these finite sample sets, is compared to the exact free energy difference. The MSE, Eq. (4.5), is then calculated by averaging over one million of such realizations. With this procedure, different combinations of overlap K, numbers of states N and sample points n are considered.

We compare three variants. Firstly, using VI, Eqs. (4.2) and (4.3), with FEP, i.e., the scheme in Fig. 4.1(a). Here, the estimates to both adjacent states are based on separate sample sets and, therefore, not correlated. Secondly, also using VI, but now with cFEP, shown in Fig. 4.1(b). In contrast to variant 1, these estimates are based on the same sample sets and, therefore, correlated. In order to keep the total computational effort constant, the number of sample points per set (i.e., per yellow point in Fig. 4.1) is two times larger for cFEP than for FEP. Thirdly, using cVI, Eqs. (4.15) and (4.16), that accounts for these correlations, also with cFEP

4.5 Results

For $N = 3$ states, Fig. 4.3(a) shows the MSEs of the three variants for different numbers of sample points. Here, for the quartic end state, $x_0 = 0$, corresponding to $K = 0.85$, was used. The corresponding configuration space densities of VI and cVI are shown in the upper row of Fig. 4.2(a).

As can be seen, cVI with cFEP, shown by the dark blue (lower) line, yields the best MSE for all numbers of sample points except very few ones. The other two variants, i.e., VI with FEP (dashed green line) and cFEP (red, upper line) yield very similar MSEs. As such, the gain in information from evaluating the Hamiltonians to both adjacent states for all sample points yields only a very small improvement compared to using separate sample sets for this purpose.

In order to quantify the improvement of cVI compared to VI for cFEP, Fig. 4.3(b) shows the ratio of the MSEs of the two variants, again in relation to the number of sample points per set. The dark orange (upper) curve ($K = 0.85$), corresponds to the MSEs shown in (a) (i.e., the values of the red curve divided by the blue curve). The improvement in the MSE plateaus slightly above two for more than two hundred sample points per state. In addition, the improvements for setups with different overlap K between the end states are shown (orange to yellow). This improvement becomes smaller for smaller values of K, but the qualitative dependence on the number of sample points remains the same.

For a constant number of sample points $n = 200$ (and $n = 100$ per set for VI with FEP, shown by the dashed green line), Fig. 4.3(c) shows how the MSEs of the three variants improve with increasing K. The MSEs converge at low K, which is in agreement with the observation from Fig. 4.2(a) that the phase space densities of the intermediate state become more similar in this case.

Figure 4.3(d) shows the MSEs for $N = 7$ states. The corresponding config-

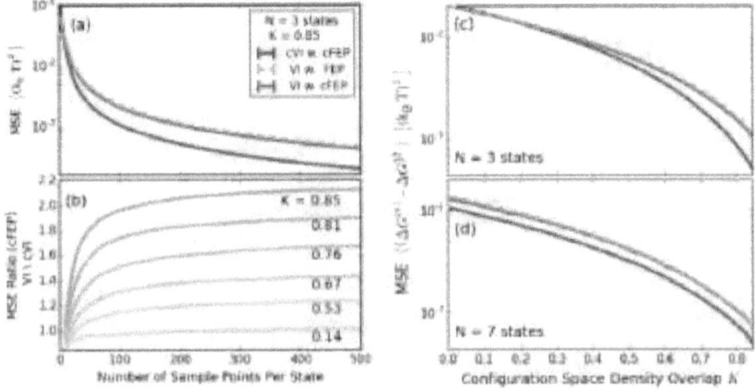

Figure 4.3: Comparison of the accuracy of VI and cVI using the schemes of Fig. 4.1. The accuracies were obtained from test simulations based on the setups shown in Fig. 4.2. (a) Using $N = 3$ states and comparing three variants of free energy calculations: Using cVI with cFEP (blue, lower solid line), VI with cFEP (red, upper solid line) and VI with FEP (green, dashed line). The MSEs of free energy calculations are shown for different number of sample points. (b) The ratio of the MSEs, and therefore, the improvement, of using cVI compared to VI for cFEP. The dark orange (upper) line ($K = 0.85$) corresponds to the ratio between the red and the blue (solid) lines in (a). In addition, the results for different configuration space density overlaps K between the end states are shown (orange to yellow). (c) Using $n = 200$ sample points, the MSEs of the three variants from (a) are shown over the full range of K. (d) As in (c), but with $N = 7$ states. The computational effort was kept constant by reducing the number of sample points per state.

uration space densities for two different values of K are shown in Fig. 4.2(b). Here, VI with FEP and cFEP still yield similar MSEs, whereas cVI with cFEP, in contrast to $N = 3$, now yields the best MSE for all K. The improvement to VI ranges from around 20 % for low K, to around 50 % for large K. This is in line with the observation from Fig. 4.2(b) that the configuration space densities between VI and cVI become more similar but do not fully converge for a larger number of states in the limit of small K.

Lastly, the cBAR estimator can be used with any choice of intermediate states for cFEP. To assess how much the cBAR estimator improves the accuracy of free energy estimates compared to BAR for cFEP, we conducted test simulations where the sampling states were chosen as in Eq. (4.1), i.e., by linear interpolation between the Hamiltonians of the end states. Test simulations were conducted at varying values of K and at $N = 5$ and $N = 7$. Evaluating the MSE, we found a statistically significant improvement, however, only in the range of $1 - 2$ % (data therefore not shown here). The improvement was independent of K and similar for both numbers of N.

Considering that the MSEs of cVI and VI can improve up to an order of magnitude compared to the linear intermediates defined in Eq. (4.1) (for a detailed comparison between VI and linear intermediates, see Ref. 230), the large majority of improvements is not due to an improved estimator, but due to the way samples are generated.

4.6 Discussion and Conclusion

In summary, we have derived a new variant of variational intermediates (cVI) that yield the optimal free energy estimate with minimal MSE when using the same sample points to evaluate the differences between the adjacent states above and below in the sequence (cFEP). This procedure is commonly used in free energy simulations, as it is computationally much cheaper to evaluate sample points at different Hamiltonians than to generate those. However, the resulting correlations between these estimates have not been considered yet.

Our test simulations for a one-dimensional Hamiltonian show that cVI with cFEP yields an improved MSE compared to the optimal sequence (VI) with FEP, i.e., using different sample points for estimators to states above and below in the sequence. For $N = 3$ states, the first variant improved the MSE by more than a factor of two for end states with high configuration space density overlap K, whereas at low K the MSEs were similar. For $N = 7$ states, the MSE improved between 20 % (low K) and 50 % (large K).

Interestingly, due to the correlations mentioned above, using VI with FEP yields only slightly worse MSEs for all K as using VI with cFEP, even though the latter involves twice as many evaluations of Hamiltonians from adjacent states. Only for cVI, thereby accounting for these correlations, the additional gain in information translates into a marked improvement of the MSE.

Similar to most other theoretical analyses and derivations of free energy calculation methods, we also needed to assume that all sample points within each intermediate state are uncorrelated. If atomistic simulations are used for sampling, the resulting time-correlations reduce the number of essentially independent sample points. Unfortunately, for our one-dimensional systems, cVI increases barrier heights, thereby increasing correlation times. We have so far not tested our method on any complex biomolecular systems, so it is unclear if these barriers can be circumvented or what the expected increase in correlation times is. However, to avoid such correlations between sample points in atomistic simulations, usually only a small subset of all sample points is used to calculate free energy differences. Based on our findings and in contrast to common practice, we therefore recommend to use different subsets to evaluate the free energy differences to different adjacent states.

The above derivation provides an example on how optimal intermediates and estimators with minimal MSE can be derived for different types of setups based on finite sampling that may help to incorporate a variety of assumptions and models into future theoretical approaches.

4.7 Appendix A: Avoiding numerical instabilities

The divergence in Eq. (4.15) at all $\mathbf{x}$ for which

$$e^{-2H_{s-1}(\mathbf{x})} r_{s-1,s}^{-2} + e^{-2H_{s+1}(\mathbf{x})} r_{s+1,s}^{-2}$$
$$= 2e^{-H_{s-1}(\mathbf{x})-H_{s+1}(\mathbf{x})} r_{s-1,s}^{-1} r_{s+1,s}^{-1} \tag{4.23}$$

causes numerical instabilities in solving the system of Eqs. (4.15) and (4.16). Replacing the factor 2 in Eq. (4.15) in the logarithm with a factor κ, i.e., for s even,

$$H_s(\mathbf{x}) = -\frac{1}{2} \ln \left(e^{-2H_{s-1}(\mathbf{x})} r_{s-1,s}^{-2} + e^{-2H_{s+1}(\mathbf{x})} r_{s+1,s}^{-2} \right.$$
$$\left. - \kappa e^{-H_{s-1}(\mathbf{x})-H_{s+1}(\mathbf{x})} r_{s-1,s}^{-1} r_{s+1,s}^{-1} \right), \tag{4.24}$$

and setting, e.g., $\kappa = 1.95$, avoids these complications. As can be easily validated, the inside of the logarithm in Eq. (4.24) is larger than zero for $0 < \kappa < 2$ for all $H_{s-1}(\mathbf{x})$ and $H_{s+1}(\mathbf{x})$. As shown for eVl in Fig. 4.2(b), $\kappa < 2$ prevents $p_s(\mathbf{x})$ to go to zero at the crossing points of $p_{s-1}(\mathbf{x})$ and $p_{s+1}(\mathbf{x})$ of the neighboring states, but is still lowered at these points.

End of publication

4.8 FurtherInterpretation

The optimal form of a cVI intermediate sampling state, Eq. 4.15, is determined through

$$e^{-2H_s(\mathbf{x})} = e^{-2H_{s-1}(\mathbf{x})} r_{s-1,s}^{-2} + e^{-2H_{s+1}(\mathbf{x})} r_{s+1,s}^{-2}$$
$$- 2e^{-H_{s-1}(\mathbf{x}) - H_{s+1}(\mathbf{x})} r_{s-1,s}^{-1} r_{s+1,s}^{-1} \tag{4.25}$$

$$\alpha \left(\frac{e^{-H_{s-1}(\mathbf{x})}}{Z_{s-1}} - \frac{e^{-H_{s+1}(\mathbf{x})}}{Z_{s+1}} \right)^2 . \tag{4.26}$$

Expressing Eq. 4.26 through configuration space densities,

$$p_s(\mathbf{x}) \ \alpha \ |p_{s-1}(\mathbf{x}) - p_{s+1}(\mathbf{x})| \ , \tag{4.27}$$

where the equality holds once the cVI sequence has converged through an iterative procedure such as the FPI described in section 3.10.

When considering only one intermediate sampling state I between the end states A and B, then Eq. 4.27 reads

$$p_I(\mathbf{x}) \ \alpha \ |p_A(\mathbf{x}) - p_B(\mathbf{x})| \ . \tag{4.28}$$

This form shows explicitly what has been observed in the context of Fig. 4.2: It is optimal to sample the regions where the configuration space densities of the end states differ. If there is no overlap, then the resulting form equals the one from VI. If the densities $p_A(\mathbf{x})$ and $p_B(\mathbf{x})$ are identical for all $\mathbf{x}$, then $H_s(\mathbf{x})$ diverges everywhere. Essentially, no sampling is required in this case.

This finding agrees with the analogy of dart board sampling, developed in the introduction in the context of Fig. 1.1(c). Here importance sampling improves the efficiency of determining the difference in area by conducting sampling only in the regions where the shapes differ. However, this result contradicts previous assumptions that most of the sampling should be conducted in the overlap region. In practice, by sampling in states from the linear interpolation scheme that shift the intermediate configuration space density from A to B, most of the sampling is conducted either in the overlap region or in regions that are relevant to neither A nor B. Whereas this practice avoids barriers in the free energy landscape, it is suboptimal for independent sampling with cFEP.

5

GROMACS Implementation

The following chapter consists of the manuscript

M. Reinhardt, H. Grubmüller, "GROMACS Implementation
for Free Energy Calculations with Non-Pairwise Variationally
Derived Intermediates".

that is currently under review in *Computer Physics Communications* and is available as a preprint at https://arxiv.org/abs/2010.14193.

Both authors contributed to conceiving the study and writing the manuscript. I conducted the described implementation and simulations.

5.1   Program Version Summary

Program Title: GROMACS-VI

CPC Library link to program files: (to be added by Technical Editor)

Developer's repository link: https://www.mpibpc.mpg.de/gromacs-vi-extension and https://gitlab.gwdg.de/martin.reinhardt/gromacs-vi-extension

Licensing provisions: LGPL

Programming language: C++14,CUDA

Supplementary material: All topologies and input parameter files required to repro-duce the example cases in this work, as well as user and developer documentation will be provided online together with the source code.

Journal reference of previous version: M.J. Abraham, T. Murtola, R. Schulz, S. Pall, J.C. Smith, B. Hess, E. Lindahl, GROMACS: High performance molecular simulations through multi-level parallelism from laptops to supercomputers, *SoftwareX* 1-2 (2015)

Does the new version supersede the previous version?: No

Reasons for the new version: Implementation of variationally derived intermediates for free energy calculations

Nature of problem: The free energy difference between two states of a thermodynamic system is calculated using samples generated by simulations based on atomistic Hamiltonians. Due to the high dimensionality of many applications as in, e.g., biophysics, only a small part of the configuration space can be sampled. The choice of the sampling scheme critically affects the accuracy of the final free energy estimate. The challenge is, therefore, to find the optimal sampling scheme that provides best accuracy for given computational effort.

Solution method(approx. 50-250 words: Sampling is commonly conducted in intermediate states, whose Hamiltonians are defined based on the Hamiltonians of the two states of interest. Here, sampling is conducted in the variationally derived intermediates states that, under the assumption of uncorrelated sample points, yield optimal accuracy. These intermediates differ fundamentally from the common intermediates in that they are non-pairwise, i.e., the forces on a particle are only additive in the end state, whereas the total force in the intermediate states cannot be split into additivecontributionsfrom the surrounding particles.

5.2   Introduction

Thermodynamic systems are driven by free energy gradients. Hence, knowledge thereof is key to the molecular understanding of a wide range of biophysical and chemical processes, as well as to applications in the pharmaceutical [8, 244, 245] and material sciences [12, 13, 246]. Consequently, *in silico* calculations of free energies are popular in providing complementary insights to experiments or assisting the selection of chemical compounds in the early stages of drug discovery projects.

The microscopic calculation of the free energy,

$$\Delta G = -\beta^{-1} \ln Z \tag{5.1}$$

$$= -\beta^{-1} \ln \int_{-\infty}^{\infty} e^{-\beta H(\mathbf{x})} d\mathbf{x} \, , \tag{5.2}$$

requires integration over all positions $\mathbf{x}$ of all particles in the system, where Z denotes the partition sum, $\beta = 1/(k_B T)$ the thermodynamic β, k_B the Boltzmann constant, T the temperature and $H(\mathbf{x})$ the Hamiltonian. As an exact integration is not feasible for high-dimensional $\mathbf{x}$ in case of many particles, sampling based approaches such as Monte-Carlo (MC) or Molecular Dynamics (MD) simulations are commonly used. Furthermore, in practice, it oftentimes suffices to know only the free energy difference between two states, which can be calculated much more accurately. The most basic approach,

$$\Delta G_{A,B} = -\beta^{-1} \ln \left\langle e^{-\beta[H_B(\mathbf{x})-H_A(\mathbf{x})]} \right\rangle_A \tag{5.3}$$

rests on the Zwanzig formula [79]. The brackets $\langle\rangle_A$ indicate an ensemble average over A is calculated. More recent methods with close relations to Eq. (5.3) that use samples from both A and B are the Bennett Acceptance Ratio (BAR) and multistate BAR (MBAR) method [80, 166] methods.

For sampling based approaches, the accuracy of a free energy difference estimate between two states A and B generally improves when sampling is not only conducted in A and B, but also in intermediate states. Commonly, a mostly linear interpolation between the end state Hamiltonians $H_A(\mathbf{x})$ and $H_B(\mathbf{x})$ is used,

$$H_{lin}(\mathbf{x}, \lambda) = (1 - \lambda)H_A(\mathbf{x}, \lambda) + \lambda H_B(\mathbf{x}, \lambda) , \tag{5.4}$$

where $\lambda \in [0, 1]$ denotes the path variable. The λ dependence of the end state Hamiltonians enables the use of soft-core potentials [91–93] that avoid divergences in case of vanishing particle for, e.g., the calculation of solvation free energies (where the molecules "vanishes" from solution). A step-wise summation,

$$\Delta G_{AB} = \sum_{i=1}^{N-1} \Delta G_{i,i+1} \tag{5.5}$$

yields the total free energy difference, where N denotes the total number of states. In the sum of Eq. (5.5), $i = 1$ corresponds to state A and $i = N$ to state B, respectively. Alternatively, for many steps the difference can be calculated with Thermodynamic Integration (TI) [90],

$$\Delta G_{AB} = \int_0^1 \left\langle \frac{\partial H(\mathbf{x}, \lambda)}{\partial \lambda} \right\rangle_\lambda d\lambda . \tag{5.6}$$

Importantly, advantageous definitions of intermediate states exist that go beyond the definition of Eq. (5.4). For example, variationally derived intermediates (VI) [239, 240] minimize the mean squared error (MSE) of free energy estimates using FEP and BAR. An easily parallelizable approximation for a small number of states is

$$H_{VI}(\mathbf{x}, \lambda) = -\tfrac{1}{2\beta} \ln \left\{ (1 - \lambda) \exp \left[-2\beta H_A(\mathbf{x}) \right] + \lambda \exp \left[-2\beta \left(H_B(\mathbf{x}) - C \right) \right] \right\},$$

(5.7)

where, similar to BAR, the free energy difference estimate is optimal if $C \approx \Delta G$. It is similar in shape to the minimum variance path (MVP) [78, 104, 105] for TI (2 vs 1/2 in the exponents). Enveloping Distribution Sampling (EDS) [100, 101], and extensions such as Accelerated EDS [102, 247] use a reference potential similar in shape to Eq. (5.7) to calculate the free energy difference between two or more end states.

Note a particular characteristic of the VI sequence and related methods, which is illustrated in Fig. 5.1: Its Hamiltonians cannot be formulated as the pair-wise sum of interaction potentials for all particles. To see this, consider the force on particle j (blue), obtained through the derivative of Eq. (5.7). It still depends on the full Hamiltonians of the end states. The consequence can be understood by considering a particle i (red), with λ dependent parameters, positioned at a distance r_{ij} so large such that all direct interactions between i and j are negligible. However, when particle i changes its position with respect to its neighboring particles, the end states Hamiltonians also change, and, therefore, so does the force on particle j.

In this work, we, firstly, describe our implementation of the VI approach, and, by extension, also the MVP and basic principles of the EDS methods for two end states, into GROMACS [157–159]. It is among the most widely used MD software packages; however, none of the above approaches are available so far in GROMACS. Secondly, we introduce an approach to avoid singularities for vanishing particles with VI.

5.3   Avoiding End State Singularities

Interestingly, the VI sequence, Eq.(5.7), already exhibits soft-core characteristics for vanishing particles, as shown in Fig. (5.2)(a) on the example of a two-particle

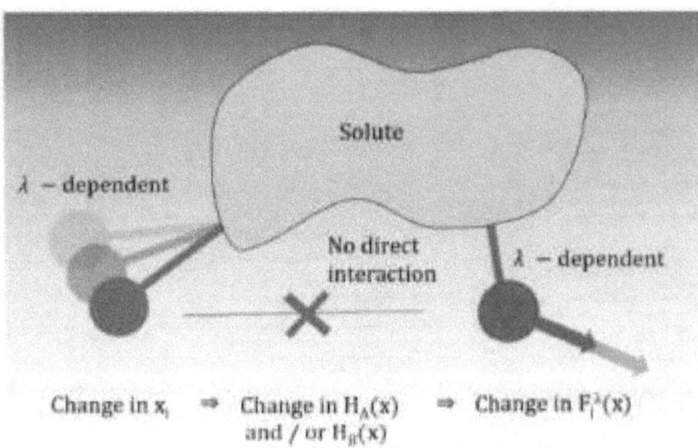

Figure 5.1: Non-pairwise potentials and forces in VI intermediates. Two particles i and j (red and blue, respectively) are considered that are λ dependent, i.e., their interaction potential differs between A and B. It is assumed that direct interactions between i and j in both A and B are negligible. If particle i changes its position, then $H_A(x)$ and / or $H_B(x)$ change accordingly, and so does $H_{VI}(x, \lambda)$. Due to the form of the VI sequence, the derivative, and therefore, the force on particle j changes.

Lennard-Jones (LJ) potential. However, divergences can still occur when configurations from the decoupled states are evaluated at foreign states, i.e., the ones that no sampling is conducted in, but that the Hamiltonian is evaluated at such as, e.g., state B in Eq. (5.3). Furthermore, when two particles start to overlap, very small changes in their separation r lead to large changes in force, which causes instabilities due to finite integration steps.

To avoid these divergences, a dependence of the end state Hamiltonians on λ analogous to common soft-core potentials [91] is introduced, i.e., $H_A = H_A(x, \lambda)$ with $H_A(x, 0) = H_A(x)$, and $H_B = H_B(x, \lambda)$, with $H_B(x, 1) = H_B(x)$. For two particle i and j with distance r_{ij}, the Coulomb and Lennard-Jones interactions in state A and B are calculated based on the modified distances r_A and r_B, respectively, that are defined as

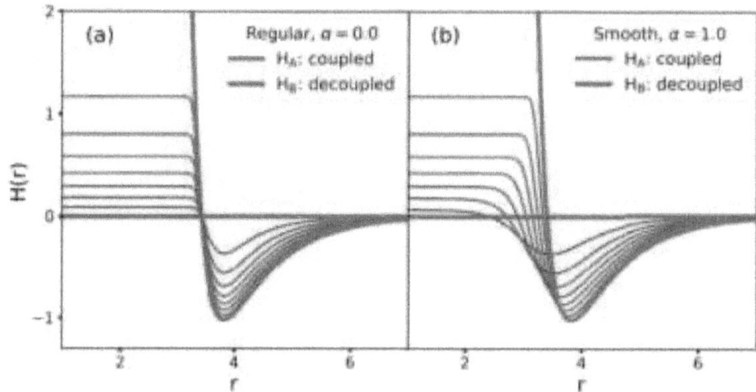

Figure 5.2: Intermediate VI states for a vanishing particle system. The thick red line shows the Lennard-Jones potential between two particles. The blue one shows the decoupled end state, i.e., the particles don't "see" each other anymore. The interpolated colors represent the intermediate states. (a) The VI sequence without and (b) with λ dependent end states.

$$r_A(r_{ij}, \lambda) = \left(\alpha\sigma_{ij}^6 \lambda^p + r_{ij}^6\right)^{\frac{1}{6}}, \tag{5.8}$$

$$r_B(r_{ij}, \lambda) = \left(\alpha\sigma_{ij}^6 (1 - \lambda)^p + r_{ij}^6\right)^{\frac{1}{6}}, \tag{5.9}$$

where α and p are soft-core parameters to be specified by the user, and σ_{ij} the Lennard-Jones parameter in the coupled state. For a system of two Lennard-Jones particles, Fig. (5.2) shows the resulting VI states without (a) and with (b) the use of λ dependent end states. As can be seen, the transition to the overlap region becomes markedly smoother.

Secondly, for increasingly complex molecules, the likelihood of barriers between the relevant parts of configuration space of the end states rises. Aside of additional techniques such as replica exchange, or meta-dynamics, the factor 2 in the exponent can be replaced by a user specific smoothing factor s introduced in the EDS [100, 101] method. In the limit of small s, a series expansion of the exponential terms yields the conventional pathway, i.e., Eq. (5.4). The modified VI sequence thus reads as

$$H_{VI}(\mathbf{x},\lambda) = -\frac{1}{s\beta} \ln \left\{ (1-\lambda)\exp\left[-s\beta H_A(\mathbf{x},\lambda)\right] \right. $$
$$\left. + \lambda\exp\left[-s\beta\Big(H_B(\mathbf{x},\lambda) - C\Big)\right] \right\} . \tag{5.10}$$

The force on particle i,

$$\mathbf{F}_i^{VI}(\mathbf{x},\lambda) = -\frac{\partial H_{VI}(\mathbf{x},\lambda)}{\partial \mathbf{x}_i} \tag{5.11}$$
$$= \exp\left[s\beta H_{VI}(\mathbf{x},\lambda)\right]$$
$$\left\{ (1-\lambda)\exp\left[-s\beta H_A(\mathbf{x},\lambda)\right] \mathbf{F}_i^A(\mathbf{x}) \right. \tag{5.12}$$
$$\left. + \lambda\exp\left[-s\beta(H_B(\mathbf{x},\lambda) - C)\right] \mathbf{F}_i^B(\mathbf{x}) \right\} ,$$

in the intermediate state characterized by λ, depends on both $H_A(\mathbf{x},\lambda)$ and $H_B(\mathbf{x},\lambda)$, as well as on the sum of the forces, $\mathbf{F}_i^A(\mathbf{x})$ and $\mathbf{F}_i^B(\mathbf{x})$ on particle i in end state A and B, respectively.

Along similar lines, the derivate

$$\frac{\partial H_{VI}(\mathbf{x},\lambda)}{\partial \lambda} = \frac{\exp\left[s\beta H_{VI}(\mathbf{x},\lambda)\right]}{\beta s}$$
$$\left\{ \left((1-\lambda)s\beta\frac{\partial H_A(\mathbf{x},\lambda)}{\partial \lambda} + 1 \right)\exp\left[-s\beta H_A(\mathbf{x},\lambda)\right] \right. \tag{5.13}$$
$$\left. + \left(\lambda s\beta\frac{\partial H_B(\mathbf{x},\lambda)}{\partial \lambda} - 1 \right)\exp\left[-s\beta(H_B(\mathbf{x},\lambda) - C)\right] \right\}$$

depends on the derivatives $\partial H_A(\mathbf{x},\lambda)/\partial \lambda$ and $\partial H_B(\mathbf{x},\lambda)/\partial \lambda$ in the end states. Equation (5.13) is used for TI.

Due to the dependence of Eq. (5.10) on C, where the accuracy is optimal if $C \approx \Delta G_{AB}$, the free energy difference has to be determined in an iterative process,

$$C^{n+1} = \Delta G_{AB'} + C^n , \tag{5.14}$$

where C^n denotes the free energy guess at iteration step n. The free energy difference $\Delta G_{AB'}$ is obtained from simulations between state A and B', where the latter denotes the end state shifted by the constant C, i.e., that is governed by

$H'_B(\mathbf{x}, \lambda) = H_B(\mathbf{x}, \lambda) - C$. The difference $\Delta G_{AB'}$ converges to zero, such that the desired quantity $\Delta G_{AB} = \Delta G_{AB'} + C^n \approx C^n$ at the end of the iteration process.

5.4 Program Structure and Usage

The end states Hamiltonians,

$$H_A(\mathbf{x}, \lambda) = H_A^\lambda(\mathbf{x}, \lambda) + H^c(\mathbf{x}) \qquad (5.15)$$

$$H_B(\mathbf{x}, \lambda) = H_B^\lambda(\mathbf{x}, \lambda) + H^c(\mathbf{x}), \qquad (5.16)$$

can be split into the λ-dependent energy contributions $H_A^\lambda(\mathbf{x}, \lambda)$ and $H_B^\lambda(\mathbf{x}, \lambda)$, and the common contributions summarized by $H^c(\mathbf{x})$ that are equal in both end states, such as water-water interactions. To calculate $H_A(\mathbf{x}, \lambda)$ and $H_B(\mathbf{x}, \lambda)$, GROMACS only evaluates the λ-dependent contributions separately for the end states, whereas $H^c(\mathbf{x})$ is calculated only once. Note that, due to the λ dependence of the end states, $H_A^\lambda(\mathbf{x}, \lambda)$ and $H_B^\lambda(\mathbf{x}, \lambda)$ differ for different intermediates for $\alpha > 0$.

The same holds for the VI sequence, Eq. (5.10). Inserting Eqs. 5.15 and 5.16, yields

$$H_{VI}(\mathbf{x}, \lambda) = H_{VI}^\lambda(\mathbf{x}, \lambda) + H^c(\mathbf{x}), \qquad (5.17)$$

where $H_{VI}^\lambda(\mathbf{x}, \lambda)$ is described by Eq. (5.10), where the end states Hamiltonians $H_A(\mathbf{x}, \lambda)$ and $H_B(\mathbf{x}, \lambda)$ have been replaced by the parts $H_A^\lambda(\mathbf{x}, \lambda)$ and $H_B^\lambda(\mathbf{x}, \lambda)$, respectively, that only sum over λ-dependent interactions. The same principle applies to the calculation of the forces and λ-derivatives. Therefore, the computational effort of VI is very close to the using conventional intermediates.

However, in the current GROMACS implementation structure, all force and energy contributions from different interaction types are interpolated between the end states right after they have been calculated, i.e., the overall calculation has the form,

$$H_{lin}^\lambda(\mathbf{x}, \lambda) = \sum_{\substack{interaction \\ type\ k}} \dots \sum_{\substack{particles \\ i,j}} (1 - \lambda)H_A^k(\mathbf{x}_{i,j}, \lambda) + \lambda H_B^k(\mathbf{x}_{i,j}, \lambda) \qquad (5.18)$$

$$F_\lambda^i(\mathbf{x}) = \sum_{\substack{interaction \\ type\ k}} \dots \sum_{\substack{particles\ j}} (1 - \lambda)F_A^k(\mathbf{x}_{i,j}, \lambda) + \lambda F_B^k(\mathbf{x}_{i,j}, \lambda). \qquad (5.19)$$

Whereas this has the least memory requirement, for VI, the full Hamiltonians and forces in the end states need to be known before the individual forces can be calculated. Therefore, the end states Hamiltonians and forces are stored separately. After all λ-dependent contributions have been collected, first the Hamiltonian and subsequently the forces are calculated.

The implementation was built based on the GROMACS 2020 version 1 (last merged with the master branch of the developer's repository on October 19th, 2019). VI can be used with the new following entries in the mdp (i.e., input parameter) file:

```
variational-morphing  = 1
smoothing-factor      = 2.
deltag-estimate       = 10.3   ; in kJ / mol
```

Furthermore, the option

```
nstcalcenergy             = 1
```

should be set, as the force calculation requires the Hamiltonians of the end state. The λ dependence of the end state Hamiltonians for VI are controlled via the already existing soft-core infrastructure,

```
sc-alpha              = 0.7
sc-r-power            = 6
sc-coul              = no
sc-sigma             = 0.3
```

By nature of Eq. 5.10, the transformation only takes place along a single λ variable, to be specified by the mdp parameter `fep-lambdas`. As such, it is not possible to decouple several interactions simultaneously with different λ spacing for each type. It is, of course, possible to decouple electrostatic and LJ interactions in a sequence, that can be defined via `coul-lambdas` and `vdw-lambdas`, respectively, whereas the other is set to either zero (full interaction) or one (no interaction) for all intermediate states.

5.5   Example and test cases

When VI is switched off, all interactions are calculated as in Eqs. (5.18), (5.19) and (5.13). To test that VI collects all contributions correctly, for the following options in the mdp file,

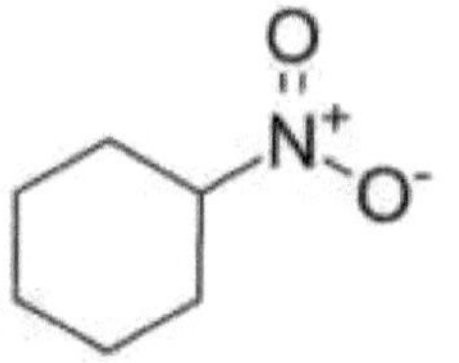

Figure 5.3: Structure of nitrocyclohexane, which is used as an example case.

```
variational-morphing   = 1
linear-test            = 1
```

Gromacs-VI calculates the intermediate Hamiltonian based on,

$$
H_{VI}^{\lambda}(\mathbf{x}, \lambda) = (1 - \lambda) \underbrace{\sum_{\substack{interaction \\ type\ k}} \cdots \sum_{\substack{particles \\ i,j}} H_A^k(\mathbf{x}_{i,j}, \lambda)}_{H_A^{\lambda}(\mathbf{x},\lambda)}
$$
$$
+ \lambda \underbrace{\sum_{\substack{interaction \\ type\ k}} \cdots \sum_{\substack{particles \\ i,j}} H_B^k(\mathbf{x}_{i,j}, \lambda)}_{H_B^{\lambda}(\mathbf{x},\lambda)} ,
\tag{5.20}
$$

and likewise, for the forces and λ derivatives. Setting the seed to a fixed value such as,

```
ld-seed   =   1
```

it can be validated that all energies required for the free energy calculation that are stored in the dhdl.xvg file match between the implementation of the VI and the conventional sequence.

Equilibrium States

As an example case, the solvation free energy of nitrocyclohexane in water was calculated (structure shown in Fig. 5.3). The topologies of the solvation toolkit package [222] created with the Generalized AMBER Force Field [221] were used. Upon energy minimization, 2 ns NVT (constant volume and temperature) and 4 ns NPT (constant pressure and temperature) equilibration were conducted, followed by 100 ns production runs.

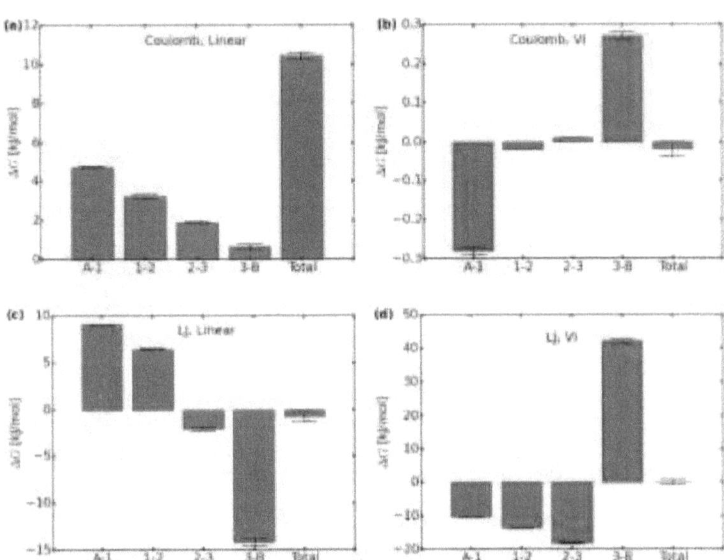

Figure 5.4: Free energy differences along intermediate states between A (coupled state) and B (decoupled state). The bars show the differences between the states denoted below. The conventional linear interpolation method, panels (a) and (c), is shown in red, whereas VI is shown in blue (panels (b) and (d)). Coulomb interactions were decoupled first (with LJ interactions still turned on), LJ interactions second (Coulomb interactions switched off).

To asses whether the VI implementation yields accurate results consistent with the ones from conventional intermediates, first, through extensive sampling with 101 states (i.e., λ steps of 0.01), a reference value value of (9.85 ± 0.02) kJ/mol was obtained. It can be divided into (10.46 ± 0.01) kJ/mol electrostatic, and (-0.61 ± 0.02) kJ/mol LJ contributions. Next, a set of simulations with 5 states, i.e., λ steps of 0.25, were conducted.

The distribution of the free energy estimates between the different states is shown for Coulomb and LJ interactions in Fig. (5.4) and differs considerably between the two methods. The bars denote the free energy difference between the states denoted at the bottom. Again, A represents the coupled, and B the decoupled state. The plots shown for VI were created based on the runs where C was set to the respective reference value, and, as such, sum up to about zero. When decoupling Coulomb interactions with a conventional linear interpolation method, shown in panel (a), the largest differences between the states occur in the first steps and gradually decreases. For VI (b), the free energy path along the intermediates has be become very small (note the differing unis on the axis). In contrast, for LJ interactions, the differences for VI (d) become larger than for the linear interpolation (c). The reason is, most likely, that the differences in the contributions from the attractive and the repulsive part of the LJ potential don't cancel for all intermediates.

To compare the accuracy of both methods, Fig. 5.5 shows the MSEs with total simulation time, distributed equally over all five states. The MSEs were obtained by dividing the trajectories of the production runs into smaller ones, and comparing the resulting free energy difference to the reference value. For VI, two different smoothing values were considered (blue and green lines), as well as an exact initial estimate (solid line) and one that is 1 kJ/mol too low (dashed lines).

For electrostatic interactions, the MSEs in Fig. 5.5(a) are significantly better for VI with $s = 2$ and an estimate close to the exact one than the MSE obtained with linear intermediates, thereby validating the result of Ref. 230. However, in this case the MSEs are quite sensitive to the initial guess. For Lennard-Jones interactions, Fig. 5.5(b), VI and linear intermediates yield similar MSEs, but the VI estimates are less sensitive to the initial guess. In both cases, the MSEs corresponding to VI with a smoothing factor of 0.1 are close to the linear ones and insensitive to the initial guess for most of the trajectory lengths in Fig. 5.5. As such, it is advantageous to start the iteration process with a smaller smoothing factor that is gradually increased with an improved estimate for C.

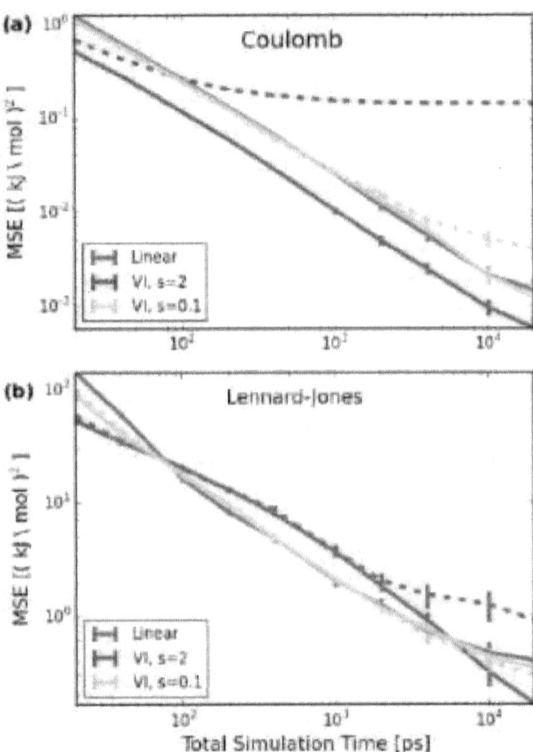

Figure 5.5: MSEs as a function of simulation time for decoupling (a) Coulomb and (b) Lennard-Jones interactions. The red line indicates the use of the conventional linear interpolation method, the blue and green line the VI approach, Eq. 5.10, using two different s values. The solid line indicate the MSEs that were obtained by using an exact initial guess, whereas a guess of 1 kJ/mol is indicated by the dashed lines.

5.6 Summary

We have implemented the VI sequence of states into the GROMACS MD software package. For Coulomb interactions, our implementations yields significantly smaller MSEs and, in this sense, higher accuracy as compared to linearly interpolated intermediates. This results requires a sufficiently accurate initial estimate, which for the test cases presented here requires only a few percent of the overall simulation time. Furthermore, using the λ dependence of the end states added to VI, for LJ interactions, similar MSEs as for conventional soft-core approaches are achieved. Given the many stepwise improvements that eventually led to the accuracy of current soft-core protocols, the fact the VI approach achieves similar accuracy already in the first attempt suggests that future refinements, e.g., of the lambda dependency on the end states, will push the accuracy even further.

5.7 Code and Data Availability

The source code is available at https://www.mpibpc.mpg.de/gromacs-vi-extension or https://gitlab.gwdg.de/martin.reinhardt/gromacs-vi-extension. Documentation, topologies and input parameter files of the above test cases are also available on the website and the repository. In the gitlab repository, all changes with respect to the official underlying GROMACS code can be retraced.

As installation is identical to that of GROMACS 2020, refer to http://manual.gromacs.org/documentation/2020/install-guide/index.html for detailed instructions.

6

LJ Analysis and Non-equilibrium Application

In the last chapter, the MSEs of VI were assessed for calculating the solvation free energy of nitrocyclohexane (see Fig. 5.5). For the decoupling of electrostatic interactions (i.e., turning all Coulomb interaction energies of nitrocyclohexane with its environment to zero), VI yielded lower MSEs than linearly interpolated intermediate states. However, for decoupling LJ interactions, the MSEs were only similar, and in some cases, even slightly worse for VI compared to established soft-core variants of the linear interpolation scheme. Therefore, the first section of this chapter will investigate the underlying reasons and identify potential approaches that could improve the accuracy of VI.

Furthermore, is has been shown empirically that if a variant of the linear interpolation scheme yields accurate predictions with equilibrium methods, then the underlying path also yields accurate predictions with non-equilibrium alchemical approaches [85, 248]. Therefore, in the second section of this chapter, it will be investigated if this finding also holds true for VI.

6.1 Separate Decoupling of vdW Attraction and Pauli Repulsion

The LJ potential,

$$V_{LJ}(r_{ij}) = \frac{C_{12}}{r_{ij}^{12}} - \frac{C_6}{r_{ij}^6} \; , \qquad (6.1)$$

expressed through the LJ parameters $C_{12} = 4\epsilon_{ij}\sigma_{ij}^{12}$ and $C_6 = 4\epsilon_{ij}\sigma_{ij}^6$ from Eq. 2.72, combines attractive (vdW) and repulsive (Pauli) interactions. However,

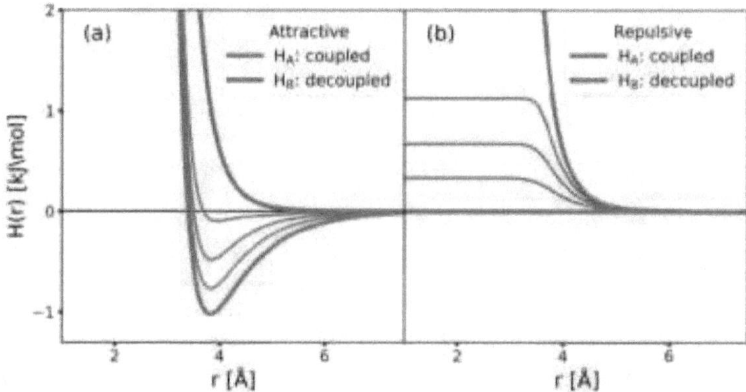

Figure 6.1: Separate decoupling of LJ interactions using VI. The Hamiltonians of the intermediates are shown as a function of distance. For illustration, the LJ parameters of argon were used [237]. (a) In the first step, attractive interactions are decoupled by setting $C_6 = 0$ in state B, while maintaining full repulsive interactions. (b) In the second step, $C_6 = 0$ is maintained, and the repulsive interactions are decoupled by setting $C_{12} = 0$ in B.

the underlying physics of these two contributions differ. To detect complications, their decoupling was analyzed in two separate sets of simulations: In the first set, repulsive interactions are maintained, whereas the attractive ones are switched off. In the second set, also the repulsive interaction energies are removed. The Hamiltonian form of the resulting VI states is shown in Fig. 6.1.

Simulation Setup

The MSEs are determined for butanol, but otherwise along similar lines as in chapter 5. For the attractive part of the LJ interactions, C_6 was changed from its regular value in state A to $C_6 = 0$ in state B whereas C_{12} remains unchanged. A reference simulation using 51 states with regular λ spacing and the linear interpolation scheme yielded a free energy difference of 62.013 ± 0.007 kJ/mol. For the repulsive part, $C_6 = 0$ was maintained for all states, whereas C_{12} was changed from its regular value to zero. The reference simulations yielded a difference of -69.95 ± 0.02 kJ/mol.

To assess the MSEs of both VI and the linear interpolation scheme, in each case five states were used for decoupling C_6 and C_{12} (regular λ steps of 0.25). In addition, a third set of simulations was conducted as a comparison, where the LJ interactions were decoupled simultaneously (i.e., equal to the procedure of decoupling LJ interactions for nitrocyclohexane in chapter 5). To keep the computational effort at the same level, ten states were used in this case (λ steps of 0.11), and the reference result was obtained from the two above as -7.94 ± 0.02 kJ/mol. For each state, upon energy minimization, 2 ns NVT and 4 ns NPT equilibration were conducted, followed by 100 ns production runs that were split into smaller trajectories to analyze the MSEs. BAR was used to calculate the free energy difference between adjacent states. For VI, the reference values were used for the estimate C. The soft-core path and parameters by Steinbrecher et al. [93] described in the methods chapter (section 2.3) was used for the linear path, and the λ dependence of the end states described in chapter 5 was used for VI.

Results

Firstly, as shown by the red lines in Fig. 6.2, decoupling the full LJ interactions simultaneously yields MSEs for VI (solid line) that are similar to the ones from the linear scheme (dashed line), as was also observed for nitrocyclohexane in chapter 5. When decoupling the attractive contribution only (blue), then the MSEs are significantly smaller, and VI is more accurate than the linear scheme. However, for removing repulsive interactions (green), both methods yield substantially higher MSEs than before. Whereas the MSEs obtained from the linearly interpolated states still decrease with simulation time, estimates of VI are highly biased, as the MSEs barely improve with increasing simulation time.

To quantify the contribution of each step to the total free energy estimate between the states in the sequence, the individual free energy differences are shown for each method and interaction type in Fig. 6.3. Only the last bar in each histogram differs, and denotes the total free energy difference between A and B. The decoupling of the full LJ interactions is shown in the first line. For the linear path, the contributions of the individual steps gradually turn from positive to negative. For VI (blue), note again that as the initial estimate was set to the reference result, a total free energy difference of zero is expected. Here, the last step (second last bar) consists of a positive free energy difference that is larger than all other (negative) ones combined. Furthermore, the difference is much larger than for any step-wise difference of the linear scheme.

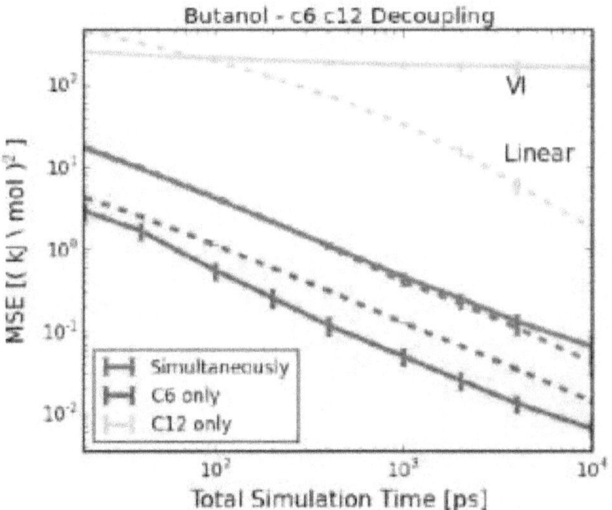

Figure 6.2: Decoupling of LJ interactions of butanol in water. The dashed lines denote that the conventional soft-core variant (Steinbrecher et al. [93]) of the linear interpolation scheme was used, the solid lines the use of the VI method for vanishing particles described in section 5.4. Three cases are considered: Firstly, the regular case of full LJ interactions in the first and none in the second end state (red). Secondly, decoupling the two contributions separately, by setting the attractive parameter from C_6 in the start to zero in the second end state, while the repulsive interactions remain (blue) and next, with all attractive interactions switched off, i.e., $C_6 = 0$, changing the repulsive term from C_{12} to zero (green).

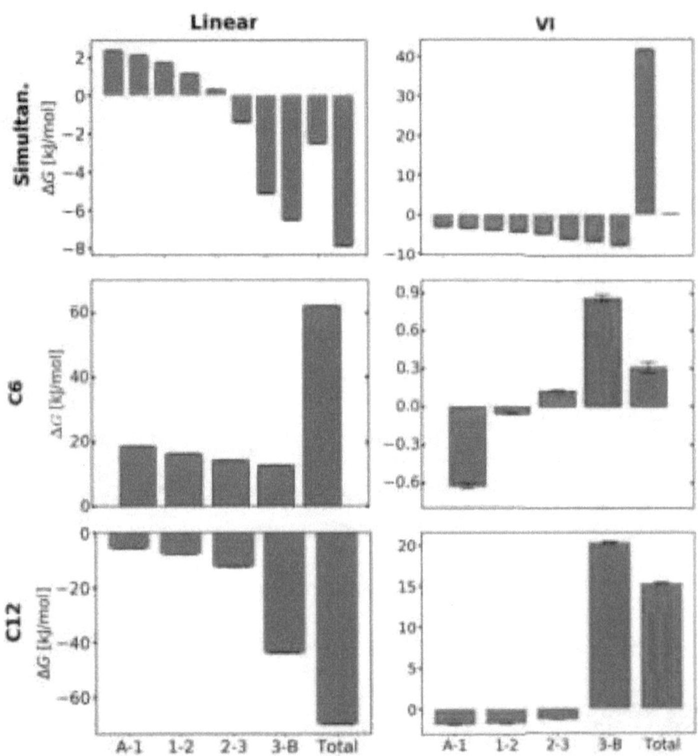

Figure 6.3: Free energy differences along intermediate states between A (coupled state) and B (decoupled state) for the three cases described for Fig. 6.2. The bars show the differences between the states denoted below: The first few bars show the difference between adjacent states, whereas the last one shows the total free energy difference between A and B. Red bars indicate that the linear interpolation scheme was used, blue bars that VI was used.

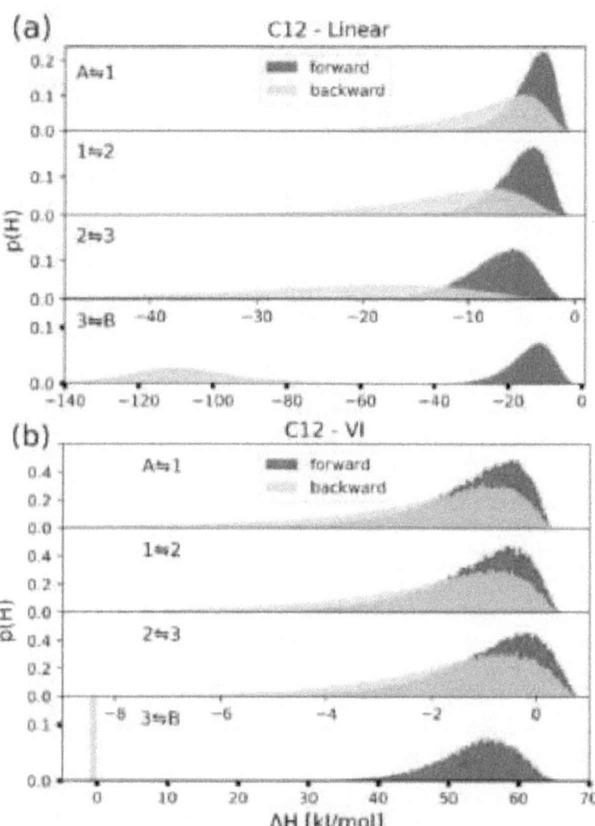

Figure 6.4: Normalized histograms of the differences in the Hamiltonians between adjacent states for the decoupling of the repulsive LJ part (butanol). $A \rightleftharpoons 1$ denotes that the forward distributions (red and blue for linear and VI, respectively) show the differences based samples from A with respect to the first intermediate, and backward *vice versa* (green). To account for the difference in direction, all backward differences have been multiplied by minus one.

For decoupling the attractive contributions (C_6), the free energy differences of all four steps are similar for the linear scheme. Along these lines, for VI, all of the differences are small. However, for the step-wise removal of repulsive interactions (C_{12}), the individual differences become very large. Both for the linear and the VI states, the largest change occurs in the last step. However, the difference to previous steps is much more drastic for VI. Strikingly, the resulting total free energy estimate is entirely wrong. It was attempted to use a λ point for state 3 that is closer to one (e.g., $\lambda = 0.99$), with the counter-intuitive result that the difference in the last step grew even larger.

Notably, some of the patterns observed for the separate decoupling resemble the ones for simultaneous decoupling. The linear scheme seems to be a combination of decoupling attractive interactions first and repulsive ones second. For VI, the large free energy differences of the last step is found both in the simultaneous and the repulsive decoupling.

To analyze the particularity of the last large step with respect to all other ones for decoupling repulsive interactions, Fig. 6.4 shows the distribution of the differences in the Hamiltonians between the individual states. For, e.g., $A \to 1$, 'forward' (red and blue) refers to the differences based on samples from state A, and 'backward' (green) to samples from state 1. The backward differences have been multiplied by minus one to account for the difference in direction. If the free energy landscapes of two states were identical (or if the Hamiltonians only differed by a constant energy offset), then the forward and (negative) backward distributions would be identical. Therefore, similar and overlapping distributions indicate similar configuration space densities (even though scenarios of disjunct configuration space densities leading to identical forward and negative backward distributions are in theory possible).

Both for the linear and the VI method, the forward and backward distributions overlap to some degree within the first three steps. Especially for VI, these distributions are very similar. In contrast, for the last step, the distributions are disjunct. For VI, the reverse distribution based on samples from the decoupled state essentially consists only of a single difference.

Discussion

Decoupling the full LJ potential yields lower MSEs than decoupling the attractive and repulsive parts in two separate steps. However, the described similarity in the histograms of step-wise free energy differences suggests that the complications observed for decoupling the repulsive part (Fig. 6.3) partly translate into decoupling the full LJ interactions, and, therefore, VI does not represent the optimum in this case.

The distributions of the differences in the Hamiltonians (Fig. 6.4) indicate that simulations in the final decoupled state take place in entirely different parts of configuration space than all other ones. However, when changing the λ value of the last intermediate close to one, transitions to the decoupled state where observed, but none going in the opposite direction. Similarly, when using starting positions from the decoupled state for all intermediates, also no transitions to the coupled state were observed.

An observation concerns the VI forces in these transition: For a vanishing particle in B, $H_B(\mathbf{x}) = 0$ and $\mathbf{F}_B(\mathbf{x}) = 0$ for all $\mathbf{x}$. Leaving the λ dependence of the end states aside, then the force, Eq. 5.12, in an intermediate state reduces to

$$\mathbf{F}_\lambda(\mathbf{x}) = \frac{(1 - \lambda)e^{-\kappa\beta H_A(\mathbf{x})}}{(1 - \lambda)e^{-\kappa\beta H_A(\mathbf{x})} + \lambda e^{\kappa\beta C}} \mathbf{F}_A(\mathbf{x}) \ . \tag{6.2}$$

In most of the cases, and for a good estimate $C \approx \Delta G \Rightarrow \langle e^{-\kappa\beta H_A(\mathbf{x})} \rangle_\lambda \approx e^{\kappa\beta C}$. Therefore, $\mathbf{F}_\lambda(\mathbf{x})$ is a non-zero fraction of $\mathbf{F}_A(\mathbf{x})$. However, if, due to fluctuations, temporarily $H_A(\mathbf{x}) >> 0$, then the force reduces to

$$\mathbf{F}_\lambda(\mathbf{x}) \approx \frac{(1 - \lambda)}{\lambda e^{\kappa\beta C}} e^{-\kappa\beta H_A(\mathbf{x})} \mathbf{F}_A(\mathbf{x}) \tag{6.3}$$

$$\alpha \ e^{-(\beta\kappa r^{-12})} r^{-13} \tag{6.4}$$

$$\text{For } r \to 0 \Rightarrow F_\lambda(\mathbf{x}) \to 0 \tag{6.5}$$

Therefore, upon a certain point, the force decreases with smaller separation r instead of increasing.

Whereas a similar phenomenon is known for conventional soft-core schemes [95], the complications are much more drastic for VI due to the characteristic of non-pairwise interactions of VI. If one atom of a vanishing molecule partly overlaps with, e.g., a water atom such that $H_A(\mathbf{x})$ is substantially increased, then not only the force $\mathbf{F}_\lambda^i(\mathbf{x})$ on atom i is decreased, but the forces of all atoms, such that the molecule becomes fully decoupled. For the molecule to go back to the coupled state, it does not suffice for one atom to be separated spontaneously by a distance of approximately σ to its neighboring atoms (as for pairwise potentials); instead, the water would have to spontaneously form a cavity for the whole solute such that no overlapping atoms remain. Naturally, this is highly unlikely for entropic reasons. Therefore, whereas transitions from the coupled to the decoupled state occur, no transitions in reverse have been observed.

Potential Approaches

The entropic problem could be avoided by maintaining a finite force for overlapping configurations that drive the system back to the coupled state. The strength of the force should be chosen such that overlapping configurations are frequently visited, but transitions back to the coupled state also occurred at a high rate.

For the linear interpolation scheme with pairwise interactions, such an approach has been developed by Gapsys et al. [95]. It was implemented into GROMACS 4.5, but not available in later versions. However, an implementation into the future GROMACS 2021 package is currently in process by Gapsys and coworkers. An alteration of this concept may subsequently be applied to VI.

In their approach, the interaction potential is defined via a switching point $r_{sw}(\lambda)$. For atom separations $r > r_{sw}(\lambda)$, the linearly interpolated interaction functions without any soft-core are used. For smaller separations, the force is linearly increased up to zero with the gradient at the switching point, as shown by the green line in Fig. 6.5(a). In contrast, the force of the most widely employed soft-core potential by Steinbrecher et al. [93] becomes zero for distances close to $r = 0$ (blue line). The potential of the soft-core interaction by Gapsys et al. [95], shown in green in Fig. 6.5(b), is determined through integration of the forces. The force and Hamiltonian are defined as

$$
\mathbf{F}(\mathbf{r}, \lambda) = \begin{cases} \mathbf{F}_{\text{lin}}(\mathbf{r}, \lambda) & , \text{ if } r \leq r_{sw}(\lambda) \\ \left.\dfrac{\mathrm{d}\mathbf{F}_{\text{lin}}(\mathbf{r}, \lambda)}{\mathrm{d}\mathbf{r}}\right|_{r = r_{sw}} (r_{sw}(\lambda) - r) + \mathbf{F}_{\text{lin}}(\frac{\mathbf{r}}{r} r_{sw}, \lambda) & , \text{ if } r < r_{sw}(\lambda) \end{cases}
\tag{6.6}
$$

$$
H(\mathbf{r}, \lambda) = \begin{cases} H_{\text{lin}}(\mathbf{r}, \lambda) & , \text{ if } r \leq r_{sw}(\lambda) \\ -\displaystyle\int_{\infty}^{r} \mathbf{F}(\mathbf{r}, \lambda)\mathrm{d}\mathbf{r} + C & , \text{ if } r < r_{sw}(\lambda) \end{cases}
\tag{6.7}
$$

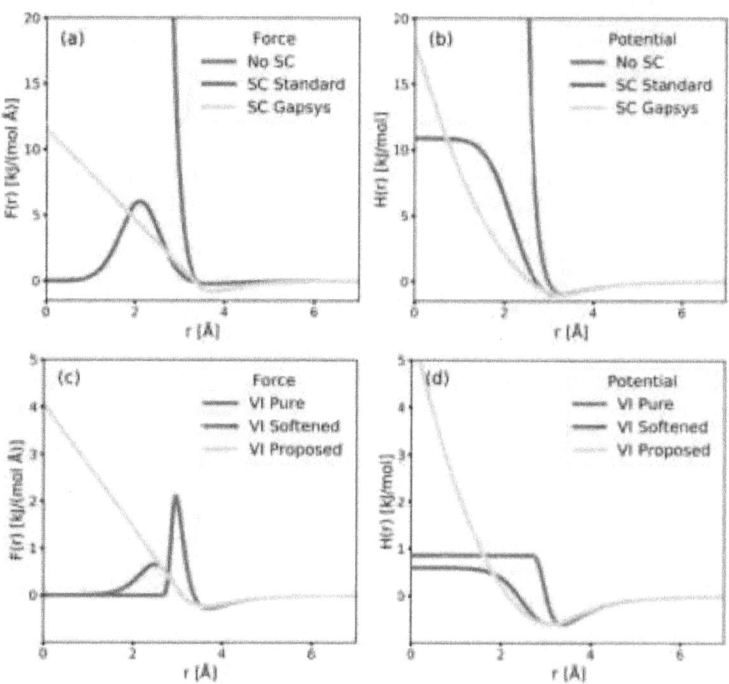

Figure 6.5: Comparison of soft-core treatments for LJ interactions in the intermediate state at $\lambda = 0.5$. Again, the LJ parameters of argon are used as an example. (a) and (b): Linear interpolation scheme. Forces and potential as a function of distance without soft-core (red), with the most widely used soft-core treatment by Steinbrecher et al. [93] (blue), and the one by Gapsys et al. [95]. (c) and (d): Forces and potential of the approximated VI sequence as derived in chapter 3 (red), of VI with the λ-dependence of the end states developed in chapter 5 with $s = 2$, $\alpha = 0.7$ (blue) and the newly proposed form (green) through adopting the approach by Gapsys et al. [95] to VI.

where C denotes the integration constant chosen such that the potential is continuous at $r = r_{sw}(\lambda)$, and $H_{\text{lin}}(\mathbf{r}, \lambda) = (1 - \lambda)H_A(\mathbf{r}) + \lambda H_B(\mathbf{r})$. The switching point is defined such that $r_{sw}(\lambda) = 0$ at both $\lambda = 0$ and $\lambda = 1$, thereby recovering the unperturbed end states.

In the one-dimensional example, this approach can be adopted by using $H_{\text{vi}}(r)$ and $F_{\text{vi}}(r)$ instead of $H_{\text{lin}}(r)$ and $F_{\text{lin}}(r)$ in Eqs. 6.6 and 6.7. The resulting forces and potentials are shown in Fig. 6.5(c) and (d). In contrast to the VI sequence derived in chapter 3 (red) and the variant with an end-state dependence on λ developed in chapter 5 (blue), a constantly increasing force is introduced for smaller r in the adopted variant (green).

For higher dimensions, a definition of the switching point via a distance criteria is not suitable for VI due its non-pairwise characteristic. However, similarly, a force can be maintained in cases where the difference in the end state Hamiltonians $\Delta H(\mathbf{x}, \lambda) = \Delta H_A(\mathbf{x}, \lambda) - H_B(\mathbf{x}, \lambda) + C$ becomes too large. Defining the switching point via an energy $E_{sw}(\lambda)$, the VI may be defined above this point, i.e., for $\Delta H(\mathbf{x}, \lambda) > E_{sw}(\lambda)$.

$$\mathbf{F}(\mathbf{r}, \lambda) = \left. \frac{d\mathbf{F}_{\text{VI}}(\mathbf{r}, \lambda, \Delta H)}{d\Delta H} \right|_{\Delta H = E_{sw}(\lambda)} \beta^{-1} \left((\beta \Delta H)^{-\frac{1}{12}} - (\beta E_{sw}(\lambda))^{-\frac{1}{13}} \right) \tag{6.8}$$
$$+ \mathbf{F}_{\text{VI}}(E_{sw}(\lambda), \lambda) \ ,$$

which, in the one-dimensional LJ cases, reduces to a force close to the one shown in green in Fig. 6.5(c) again. Such an approach could drastically reduce the entropic barrier separating the coupled from the decoupled state.

6.2   Non-Equilibrium Application

To analyze the approximated VI path for non-equilibrium transitions, the MSEs of two calculations were assessed: Firstly, of the free energy difference between an argon LJ gas, consisting of 100 atoms, and an ideal argon gas. Secondly, of the free energy difference of charged and uncharged butanol in solution.

Simulation Setup

For argon, the same LJ parameters as in section 3.6 were used ($\sigma = 3.405$ Å, $\epsilon = 1.0446$ kJ/mol and $m = 39.95$ u). In this case, the simulations were conducted with the GROMACS VI implementation described in chapter 5 and in an NPT ensemble, with $T = 298.15$ K and $P = 1.013$ bar. A reference free energy difference of 0.158 ± 0.005 kJ/mol was obtained using 101 equilibrium states with 100 ns simulation time in each one of them. For butanol, the same system setup and reference value as described in section 3.6 was used.

For both systems, 1000 non-equilibrium trajectories were simulated in each direction, i.e., going from A (coupled state) to B (decoupled state) and in reverse. For argon, an increment of $\Delta\lambda = 2 \cdot 10^{-5}$ per time step was used and for butanol $\Delta\lambda = 2 \cdot 10^{-6}$. The starting points were drawn from 100 ns equilibrium simulations in both A and B (i.e., one starting configuration every 0.1 ns). This procedure was conducted using the linear interpolation scheme, as well as for VI with smoothing values $s = 2$ and $s = 0.1$. In case of the LJ gas, soft-core was employed, with $\alpha = 0.5$ for linear, and using $\alpha = 0.7$ for the end state dependence of VI. In all setups, VI calculations were conducted using the reference value as the initial estimate. For butanol and both smoothing values, and additional set of simulations was performed were the initial estimate C was 1 kJ/mol too small.

To assess the MSEs of all paths, free energy estimates were calculated based on trajectory sets of varying sizes, ranging from 3 to 200 in each direction. For each size, multiple free energy estimates were calculated by using shifted sets of trajectories ranging from 500 sets (for 3 trajectories) to 20 sets (for 200 trajectories). Estimates of the free energy difference were calculated based on the work values from non-equilibrium trajectories as described in the methods section 2.4 using both the adapted BAR and the CGI method. In none of the above cases, the resulting MSEs differed significantly between BAR and CGI, so the results being shown in the following are the ones obtained with BAR using the implementation within the pmx tool [162].

Results

VI (solid blue line in Fig. 6.6) yields significantly higher MSEs for both calculations on (a) argon and (b) butanol than for the linear path (red). VI with smoothing ($s = 0.1$, green line) yields MSEs that are similar to the linear path. For an initial estimate that differs from the exact free energy difference (dashed lines), the MSEs are higher, and the fast flattening of the curve with the number of trajectories indicates a systematic bias.

To investigate the reasons leading to the higher MSEs of VI, the underlying work distributions for butanol are shown in Fig. 6.7. These are much wider for VI than for the linear path and for VI with $s = 0.1$. Particularly, rare but large work values in both directions occur. Furthermore, when C does not equal the reference value, in the optimal case the distributions would be equal to the one on the left but centered around a mean of 1 kJ/mol. Instead, these become non-symmetric.

On the example of a single trajectory for the linear path and VI each (C exact, $s = 2$), $\frac{\partial H}{\partial \lambda}$ is shown as a function of λ for butanol in Fig. 6.8. As can be seen, the plots highly differ between the linear and the VI method. For the linear scheme, the derivate gradually decreases with λ. For VI, the derivatives are large for λ close to zero and to one, whereas the derivatives are small for the majority of the λ range in between.

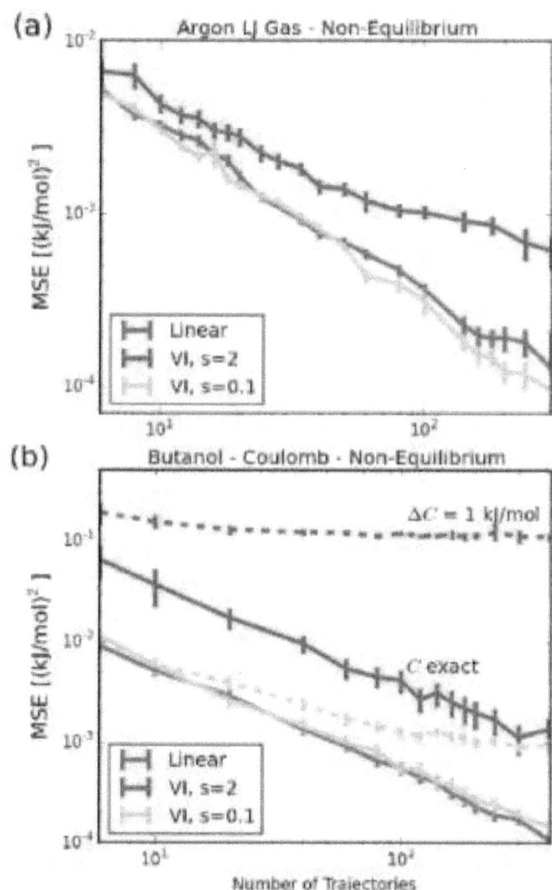

Figure 6.6: MSEs for different numbers of non-equilibrium trajectories. Half of the trajectories were conducted in the forward, and half in reverse direction. VI with different smoothing (red and green) is compared to the linear path (red) for the calculation of (a) the free energy difference between an argon LJ and an ideal gas and (b) the difference between charged and uncharged butanol in solution.

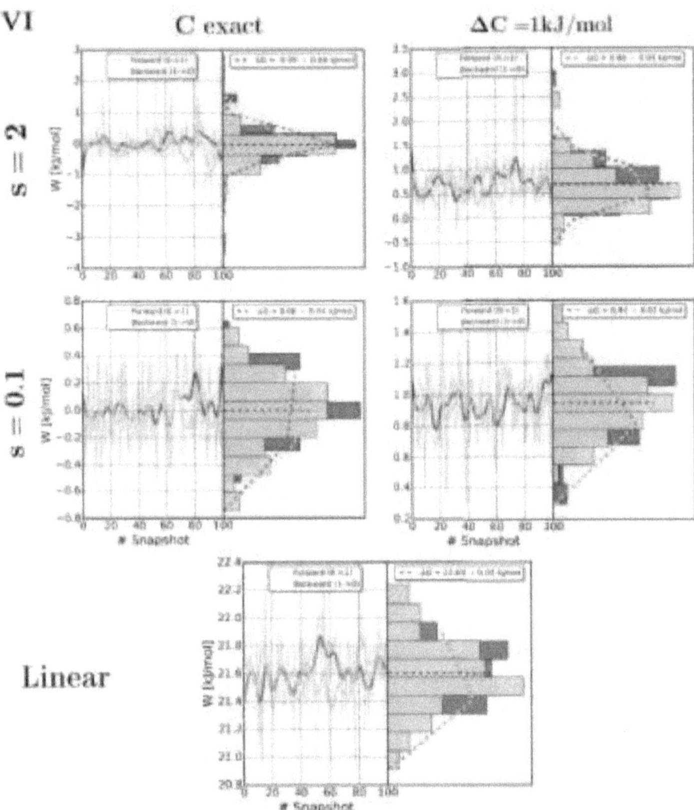

Figure 6.7: Work distributions from non-equilibrium trajectories for butanol: comparison between VI and linear path. For VI, different combinations of an initial estimate C (left: C equals the reference value, right: C is 1 kJ/mol below the reference value) and smoothing parameter (top: VI, $s = 2$, middle: VI, $s = 0.1$) are shown. Within each plot, the left half shows the work as a function of the trajectory number in slightly transparent colors for the first 100 trajectories, both in forward (blue and red) and reverse direction (green). The thicker, non-transparent line shows a moving average. The histograms in the right part combine the work values from the left part. Based on these histograms, the free energy difference was calculated, as shown in the upper right part of each plot. The images were created based on the analysis with the pmx tool [162].

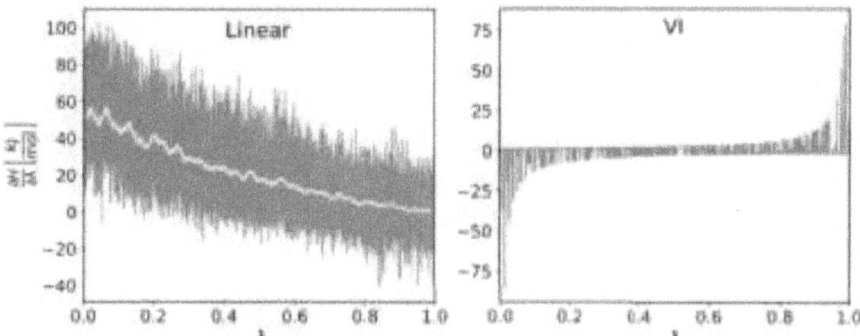

Figure 6.8: $\frac{\partial H(\mathbf{x})}{\partial \lambda}$ as a function of λ for butanol over a single trajectory in forward direction out of equilibrium. The total work is obtained through integration over λ. Left: linear path, the white line presents a moving average over a width of 0.05. Right: VI (C exact and $s = 2$).

Discussion

In its current form, VI is less accurate than linear paths for non-equilibrium applications. Several points related to the accuracy are discussed in the following:

Firstly, non-independent samples in the equilibrium simulations that the starting positions were drawn from — a common reason for inaccuracies non-equilibrium based simulations — did most likely not cause the complications observed in this section. As a start, due to its relatively flat free energy landscape, it is easier to produce samples that can be assumed as independent for an argon LJ gas than for a molecular system in solution, and the latter would therefore be more affected by such shortcomings. However, the difference in accuracy between the two methods is similar for both butanol and argon, especially at a higher number of trajectories. Furthermore, correlations are often observed through systematic trends of the work values as a function of trajectory number. Whereas the fact that no such trend was observed for the work distributions in Fig. 6.7 is not a proof that the accuracies are not affected by correlations between starting positions, it gives another indication that the complications observed in this chapter are most likely caused elsewhere.

Secondly, one potential reason for the inaccuracies of VI is related to the fact that the largest contributions to the integral over $\frac{\partial H}{\partial \lambda}$ are distributed close to $\lambda = 0$ and 1. Therefore, only a small amount of sampling is conducted in these λ regions compared to the total length of the trajectories. As the two large contributions to the integral at the end of the λ range have an opposite sign, any fluctuations therein cause large fluctuations in the total work, as observed in Fig. 6.7. Discretization through finite $\Delta\lambda$ steps will therefore more strongly translate into integration errors for VI.

Potential Approaches

Improvements in the accuracy could therefore be achieved by directing more sampling to the λ regions with the highest absolute values in the $\frac{\partial H}{\partial \lambda}$ and, to accurately integrate over the regions with changes therein, with high second derivatives $\frac{\partial^2 H}{\partial \lambda^2}$. An adaptive $\Delta\lambda$ increment that is small in these cases and high in regions with small first and second derivatives would achieve this goal.

Note that without soft-core, as discussed in the methods section 2.3, any spacing for one set of prefactors in front of the exponentials in Eq. 5.10 corresponds to a different spacing for another set of prefactors. For example, constant $\Delta\lambda$ increments for the prefactors $(1 - \lambda)^4$ and λ^4 lead to identical states as the prefactors $(1 - \lambda)$ and λ with some other form of irregular spacing. Therefore, alternatively, the form of the prefactors could be optimized. However, the advantage of an adaptive $\Delta\lambda$ increment is that the the optimal form prefactors does not have to be assumed or determined *a priori*. Nonetheless, even if sampling along λ is optimized via either option, it is as of now unclear how close the functional form of the approximated VI path itself is to the optimum for non-equilibrium methods. Ideally, the optimal functional form with prefactors that are optimal for constant spacing would be derived through an analytical derivation for non-equilibrium approaches.

7

Conclusion and Outlook

7.1 Conclusion

This thesis addressed the question which sequence of intermediate states, out of all possible ones, yields the highest accuracy. In this context, the highest accuracy refers to sampling based estimates with the smallest mean-squared error (MSE) with respect to the exact free energy difference. For two of the most widely used methods to calculate free energy differences via atomistic simulations, free energy perturbation (FEP) [79] and the Bennett acceptance ratio method (BAR) [80], this sequence was derived in chapter 3 using variational calculus, and is hence referred to as the Variationally derived Intermediates (VI) method.

The VI sequence offers a number of insights: Firstly, it shows that the form of these optimal intermediates is very different from established ones. Whereas the most widely used linear schemes, and soft-core variants thereof, interpolate the Hamiltonians of the end states A and B, VI is closer to the interpolation of the squared configuration space densities. The resulting densities decrease in the region of A and increase in the region of B. Hence, unlike previously assumed, sampling of configurations that lie between A and B in configuration space, but are relevant to neither state, is therefore not efficient and should only enable transitions between these regions. For tests on one-dimensional systems, substantial improvements were observed for VI compared to states from the linear interpolation scheme. These improvements were the most pronounced (a factor of two or more) for the notoriously difficult cases where the configuration space densities of the end states overlap only by a small amount.

Secondly, as VI is optimal for any number of states in the sequence, it is a generalization of the two main alternatives to the linear interpolation scheme. In the limit of many states, VI yields the same MSEs as the minimum variance path (MVP) [78, 104, 105], which was derived in this limit for Thermodynamic Integration (TI) [90]. In contrast, when using only one intermediate state, VI resembles the empirically constructed Enveloping Distribution Sampling (EDS) potential [100].

Thirdly, VI was directly optimized as a sequence of discrete states. As a result, it differs from all other forms developed in the past in that the intermediate states are coupled. As such, these are not directly defined as a functional of the end state Hamiltonians, but instead via adjacent states in the sequence, and the optimal solution is determined through a system of equations. Conceptually, only in reverse a path may be implied by taking the limit of infinitely many intermediates and defining a path variable via the fractional labels of these intermediates. Essentially, by optimizing the sequence of discrete states, VI not only accounts for the bias that arises from such a discretization, but also yields a lower variance than, e.g., the MVP, which is only optimal in the limit of many steps.

Fourthly, for FEP with only one virtual intermediate (i.e., in which no sampling is conducted) between the end states, VI yields the BAR formula. Importantly, via this connection VI offers a new and more general derivation of BAR, which was initially derived by optimizing the variance in the large sample limit and under the assumption that the differences in the Hamiltonians follow a Gaussian distribution. The VI derivation therefore shows that, firstly, BAR not only optimizes the variance but also the MSE, and secondly, also does so for finite numbers of sample points with less restrictive assumptions than the initial derivation. Furthermore, BAR represents an implicit problem, where the optimal estimator depends on an estimate of the free energy itself, and is therefore solved iteratively. VI generalizes the BAR principle for many states, as not only the estimator, but also the form of all intermediate states is determined through such an iterative procedure.

In the next step, we considered correlated Free Energy Perturbation (cFEP), which is an extended setup to calculate free energy differences. For cFEP, the same sample points are used to evaluate the free energy differences to adjacent states in the sequence, rather than separate ones. This procedure is common practice, as it is much more efficient to evaluate sample points with multiple Hamiltonians than to generate uncorrelated ones. However, correlations arise between the step-wise free energy estimates that had not been accounted for prior to this work. The sequence of intermediate states (cVI) yielding optimal MSEs under these conditions was derived and assessed for a number of 1-D systems, giving, in the optimal case, an improvement up to more than a factor of two compared to using VI with either FEP or cFEP. The equivalence between virtual intermediates and estimators was further used to develop an improved estimator (cBAR) for cFEP with arbitrarily chosen intermediates. However, for linearly interpolated intermediates, cBAR only yielded improvements of about one to two percent compared to regular BAR.

Interestingly, the configuration space density of a single cVI sampling state is proportional to the absolute difference of the densities from the end states. Therefore, it is optimal to sample in the regions that are only relevant to one, but not to both end states. This fact contradicts previous assumptions that intermediates should be chosen such that they sample the overlap region and, therefore, rather represents an "anti overlap" principle. As such, it is in line with the dart-board analogy from Fig. 1.1(c) of the introduction: A higher accuracy is achieved through importance sampling in the regions where the shapes differ instead of sampling in the overlapping bulk.

We next considered the application of VI and cVI to realistic atomistic many-particle systems. Fundamentally, there is no obstacle to apply the VI and cVI sequences to such systems. However, their parallelization and implementation into current software packages is nonetheless challenging due to the difference of the coupled form from established paths. To avoid this technical drawback, a sequence based on an approximated path was devised that yielded similar MSEs as the optimal VI sequence at a small number of intermediates in one-dimensional systems. The approximated VI sequence was implemented, firstly, into a self-written program that models a LJ gas, and secondly, into the GROMACS MD software package. The implementation into the latter was described in detail in chapter 5.

The approximated VI sequence yields substantial improvements for a number of systems compared to the linear interpolation scheme. To calculate the free energy difference for a change in atom type of a LJ gas, almost ten times less sampling was required to achieve the same accuracy. For the calculation of the solvation free energy difference between charged and uncharged butanol, about two times less sampling was required. Note again that, as outlined in the chapter on theory and methods (2), the MSEs were assessed by comparing the estimates from numerous short simulations to a converged reference result based on the same Hamiltonian. For comparisons with experiments, not only the sampling errors, but also force field inaccuracies would contribute to the MSEs, and therefore not allow a conclusive evaluation of our sampling method.

In chapter 5, VI was tested on one of the most challenging cases, the (de)solvation of an entire molecule (nitrocyclohexane). Interestingly, the VI sequence, as well as the approximation thereof, already exhibit soft-core properties. However, a lambda dependence of the end state Hamiltonians still had to be introduced to avoid numerical instabilities in regions of large gradients. For the calculation of the solvation free energy of nitrocyclohexane, similar MSEs were obtained between VI and the established linear soft-core approach.

In the last step, based on empirical findings that paths yielding accurate results for equilibrium methods also do so for non-equilibrium ones [85, 248], we also tested the path underlying the approximated VI sequence with a non-equilibrium approach. However, in the cases analyzed in chapter 6 — the transformations of a LJ to an ideal gas as well as from charged to uncharged butanol in solution — the subsequent estimates were less accurate than the ones of established soft-core variants of linear interpolation schemes. An analysis of the individual work distributions revealed that most of the changes in the simulated system occurred within a small fraction of the entire λ range. As of now, it remains unclear if a suboptimal distribution of sampling along λ, caused by an inappropriately constant $\Delta\lambda$ increment, is the sole reason for obtaining higher MSEs than the linear interpolation scheme, or if the approximated VI path itself is far from the optimal functional form for non-equilibrium approaches.

A major challenge remains: As most of the analytical approaches in the context of free energy calculations, all of the above presented analytical results were derived assuming independent sample points. However, in practice, this assumption is violated in MD simulations to an extent that depends on factors such as simulation time, the choice of starting points and barriers within the free energy landscape. Unfortunately, VI seems to be more affected by this sampling problem than linear interpolation schemes. In the extreme case of two completely disjunct configuration space densities of the end states, the intermediate states would even consist of two disconnected regions. In these cases, which region is explored is entirely dependent on the starting position, even for infinitely long trajectories. For cVI, even for connected regions, large barriers occur and cVI has, for this reason, so far not been tested on many-particle systems.

Whereas for the calculation of solvation free energies presented in this thesis no substantial enthalpic barriers were encountered, a detailed analysis on butanol (chapter 6) revealed that entropic barriers lead to VI yielding MSEs that were not more accurate than conventional soft-core approaches. The problem can partly be mitigated by introducing a smoothing factor in the exponents, as was introduced in the EDS method [100], that gradually switches between the linear and the exponential form of a potential. Furthermore, an approach avoiding entropic barriers through maintaining non-zero forces for overlapping particles in intermediate states has been suggested.

In summary, the VI and cVI derivations provide the sequences of intermediate states with minimal MSEs assuming independent sample points. Marked improvements in accuracy compared to linear intermediates were observed for one-dimensional test systems, and a number of fundamental insights were obtained. Furthermore, VI also provided estimates of free energy differences with improved accuracy for several many-particle systems. The major challenge, i.e., non-independent sample points, still needs to be addressed through approaches outlined in the following section.

7.2 Outlook

This work may be extended into several directions. These can be broadly categorized into, firstly, analytical approaches addressing some of the more fundamental questions in the context of VI, and secondly, future applications of the concepts from this work.

Analytical Approaches

Firstly, it has been verified empirically on a variety of test cases that the fixed point iteration used to solve the system of equations defining the optimal VI sequence converges to a unique solution, independent of the initial guess. However, attempts to analytically proof — or refute — this finding were so far unsuccessful. In case this empirical finding was incorrect, potentially a solution yielding even lower MSEs than the one analyzed in this work may exist.

Secondly, it was empirically shown that in the limit of many states the VI sequence yields the same MSEs as the MVP which, however, also remains to be proven analytically.

Thirdly, the cVI derivation may be extended for a setup in which not only the difference in enthalpy of an intermediate to the two adjacent ones is used, but instead the enthalpy difference to all other states.

Essentially, the "holy grail" of theories for free energy calculations would derive the optimal sequence of states without assuming independent sample points. It would also include correlations between subsequent points by, e.g., considering sampling as a diffusive process in a free energy landscape. It would thereby avoid complications arising from both enthalpic and entropic barriers.

Application

Firstly, the complication that regions in conformation space are separated through entropic barriers for the calculation of solvation free energies with VI needs to be addressed. To this aim, we have suggested an approach maintaining non-zero forces for overlapping vanishing particles in intermediates states, as outlined in chapter 6. This approach was adopted from a soft-core variant developed by Gapsys et al. [95] for linear interpolation schemes. As it already yielded marked improvements for linear intermediates that suffer less from this problem, improvements are also expected for VI.

Secondly, barriers in the free energy landscape may, in addition, be overcome through the combination of VI with sampling techniques devised for this purpose. Examples, briefly introduced in the introduction, include metadynamics, conformational flooding or replica exchange (RE). The latter is readily usable with the current GROMACS VI implementation, but will require optimization of the acceptance criteria to switch conformations between different states.

Thirdly, an optimization of the iteration process updating the free energy estimate C, which the approximated VI sequence depends on, needs to be conducted. An update of C is only preferable if the prediction based on the obtained data to this point has a significantly lower standard deviation than the difference to the previously used estimate. The optimal criterion, however, still needs to be determined. Furthermore, a future approach may address how to extract and combine the information from trajectories based on different C to make full use of the initial optimization runs. For example, a variance weighted average is an option, which may, however, still be biased to an unknown degree.

Fourthly, the implementation of the optimal VI sequence instead of the approximated one has already been started, but not sufficiently tested, yet. Similar to the approximated sequence, the estimated ratio of the partition sums between adjacent states is updated iteratively based on the free energy estimates obtained to this point.

Lastly, for non-equilibrium simulations, it can be determined how much the observed inaccuracies of VI result from a suboptimal distribution of sampling capacities along λ. To do so, a variable $\Delta\lambda$ increment that would be small for large first and second order derivatives $\frac{\partial H_\lambda}{\partial \lambda}$ and $\frac{\partial^2 H_\lambda}{\partial \lambda^2}$, respectively, and *vice versa*, may be designed. Such an adaptive increment would lead to better sampling in λ ranges where large changes in the system occur, whereas avoiding oversampling the ones with few changes therein. As this disparity was particularly large for VI with non-equilibrium methods, substantial improvements are expected here through such an adaptive approach, but applications based on other paths may benefit as well.